HOLT SCIENCE & TECHNOLOGY

Microorganisms, Fungi, and Plants

ANNOTATED TEACHER'S EDITION

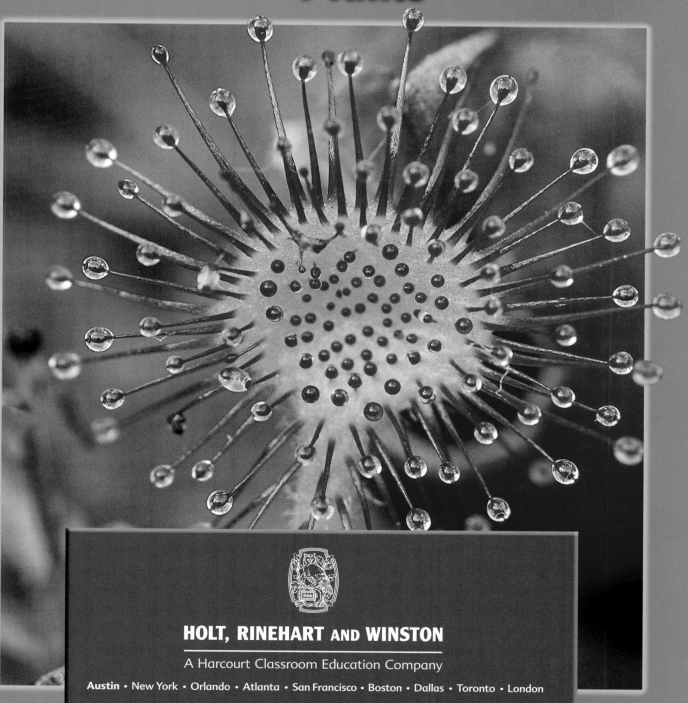

HOLT, RINEHART AND WINSTON

A Harcourt Classroom Education Company

Austin • New York • Orlando • Atlanta • San Francisco • Boston • Dallas • Toronto • London

Acknowledgments

Chapter Writers

Katy Z. Allen
Science Writer and Former Biology Teacher
Wayland, Massachusetts

Linda Ruth Berg, Ph.D.
Adjunct Professor–Natural Sciences
St. Petersburg Junior College
St. Petersburg, Florida

Jennie Dusheck
Science Writer
Santa Cruz, California

Mark F. Taylor, Ph.D.
Associate Professor of Biology
Baylor University
Waco, Texas

Lab Writers

Diana Scheidle Bartos
Science Consultant and Educator
Diana Scheidle Bartos, L.L.C.
Lakewood, Colorado

Carl Benson
General Science Teacher
Plains High School
Plains, Montana

Charlotte Blassingame
Technology Coordinator
White Station Middle School
Memphis, Tennessee

Marsha Carver
Science Teacher and Dept. Chair
McLean County High School
Calhoun, Kentucky

Kenneth E. Creese
Science Teacher
White Mountain Junior High School
Rock Springs, Wyoming

Linda Culp
Science Teacher and Dept. Chair
Thorndale High School
Thorndale, Texas

James Deaver
Science Teacher and Dept. Chair
West Point High School
West Point, Nebraska

Frank McKinney, Ph.D.
Professor of Geology
Appalachian State University
Boone, North Carolina

Alyson Mike
Science Teacher
East Valley Middle School
East Helena, Montana

C. Ford Morishita
Biology Teacher
Clackamas High School
Milwaukie, Oregon

Patricia D. Morrell, Ph.D.
Assistant Professor, School of Education
University of Portland
Portland, Oregon

Hilary C. Olson, Ph.D.
Research Associate
Institute for Geophysics
The University of Texas
Austin, Texas

James B. Pulley
Science Editor and Former Science Teacher
Liberty High School
Liberty, Missouri

Denice Lee Sandefur
Science Chairperson
Nucla High School
Nucla, Colorado

Patti Soderberg
Science Writer
The BioQUEST Curriculum Consortium
Beloit College
Beloit, Wisconsin

Phillip Vavala
Science Teacher and Dept. Chair
Salesianum School
Wilmington, Delaware

Albert C. Wartski
Biology Teacher
Chapel Hill High School
Chapel Hill, North Carolina

Lynn Marie Wartski
Science Writer and Former Science Teacher
Hillsborough, North Carolina

Ivora D. Washington
Science Teacher and Dept. Chair
Hyattsville Middle School
Washington, D.C.

Academic Reviewers

Renato J. Aguilera, Ph.D.
Associate Professor
Department of Molecular, Cell, and Developmental Biology
University of California
Los Angeles, California

David M. Armstrong, Ph.D.
Professor of Biology
Department of E.P.O. Biology
University of Colorado
Boulder, Colorado

Alissa Arp, Ph.D.
Director and Professor of Environmental Studies
Romberg Tiburon Center
San Francisco State University
Tiburon, California

Russell M. Brengelman
Professor of Physics
Morehead State University
Morehead, Kentucky

John A. Brockhaus, Ph.D.
Director of Mapping, Charting, and Geodesy Program
Department of Geography and Environmental Engineering
United States Military Academy
West Point, New York

Linda K. Butler, Ph.D.
Lecturer of Biological Sciences
The University of Texas
Austin, Texas

Barry Chernoff, Ph.D.
Associate Curator
Division of Fishes
The Field Museum of Natural History
Chicago, Illinois

Donna Greenwood Crenshaw, Ph.D.
Instructor
Department of Biology
Duke University
Durham, North Carolina

Hugh Crenshaw, Ph.D.
Assistant Professor of Zoology
Duke University
Durham, North Carolina

Joe W. Crim, Ph.D.
Professor of Biology
University of Georgia
Athens, Georgia

Peter Demmin, Ed.D.
Former Science Teacher and Chair
Amherst Central High School
Amherst, New York

Joseph L. Graves, Jr., Ph.D.
Associate Professor of Evolutionary Biology
Arizona State University West
Phoenix, Arizona

William B. Guggino, Ph.D.
Professor of Physiology and Pediatrics
The Johns Hopkins University School of Medicine
Baltimore, Maryland

David Haig, Ph.D.
Assistant Professor of Biology
Department of Organismic and Evolutionary Biology
Harvard University
Cambridge, Massachusetts

Roy W. Hann, Jr., Ph.D.
Professor of Civil Engineering
Texas A&M University
College Station, Texas

Copyright © 2002 by Holt, Rinehart and Winston

All rights reserved. No part of this publication may be reproduced or transmitted in any form or by any means, electronic or mechanical, including photocopy, recording, or any information storage and retrieval system, without permission in writing from the publisher.

Requests for permission to make copies of any part of the work should be mailed to the following address: Permissions Department, Holt, Rinehart and Winston, 10801 N. MoPac Expressway, Austin, Texas 78759.

For permission to reprint copyrighted material, grateful acknowledgment is made to the following sources:

sciLINKS is owned and provided by the National Science Teachers Association. All rights reserved.

The name of the **Smithsonian Institution** and the sunburst logo are registered trademarks of the Smithsonian Institution. The copyright in the Smithsonian Web site and Smithsonian Web site pages are owned by the Smithsonian Institution. All other material owned and provided by Holt, Rinehart and Winston under copyright appearing above.

Copyright © 2000 **CNN** and **CNNfyi.com** are trademarks of Cable News Network LP, LLLP, a Time Warner Company. All rights reserved. Copyright © 2000 Turner Learning logos are trademarks of Turner Learning, Inc., a Time Warner Company. All rights reserved.

Printed in the United States of America

ISBN 0-03-064773-8

1 2 3 4 5 6 7   048   05 04 03 02 01 00

Acknowledgments (cont.)

John E. Hoover, Ph.D.
Associate Professor of Biology
Millersville University
Millersville, Pennsylvania

Joan E. N. Hudson, Ph.D.
Associate Professor of Biological Sciences
Sam Houston State University
Huntsville, Texas

Laurie Jackson-Grusby, Ph.D.
Research Scientist and Doctoral Associate
Whitehead Institute for Biomedical Research
Massachusetts Institute of Technology
Cambridge, Massachusetts

George M. Langford, Ph.D.
Professor of Biological Sciences
Dartmouth College
Hanover, New Hampshire

Melanie C. Lewis, Ph.D.
Professor of Biology, Retired
Southwest Texas State University
San Marcos, Texas

V. Patteson Lombardi, Ph.D.
Research Assistant Professor of Biology
Department of Biology
University of Oregon
Eugene, Oregon

Glen Longley, Ph.D.
Professor of Biology and Director of the Edwards Aquifer Research Center
Southwest Texas State University
San Marcos, Texas

William F. McComas, Ph.D.
Director of the Center to Advance Science Education
University of Southern California
Los Angeles, California

LaMoine L. Motz, Ph.D.
Coordinator of Science Education
Oakland County Schools
Waterford, Michigan

Nancy Parker, Ph.D.
Associate Professor of Biology
Southern Illinois University
Edwardsville, Illinois

Barron S. Rector, Ph.D.
Associate Professor and Extension Range Specialist
Texas Agricultural Extension Service
Texas A&M University
College Station, Texas

Peter Sheridan, Ph.D.
Professor of Chemistry
Colgate University
Hamilton, New York

Miles R. Silman, Ph.D.
Assistant Professor of Biology
Wake Forest University
Winston-Salem, North Carolina

Neil Simister, Ph.D.
Associate Professor of Biology
Department of Life Sciences
Brandeis University
Waltham, Massachusetts

Lee Smith, Ph.D.
Curriculum Writer
MDL Information Systems, Inc.
San Leandro, California

Robert G. Steen, Ph.D.
Manager, Rat Genome Project
Whitehead Institute—Center for Genome Research
Massachusetts Institute of Technology
Cambridge, Massachusetts

Martin VanDyke, Ph.D.
Professor of Chemistry, Emeritus
Front Range Community College
Westminister, Colorado

E. Peter Volpe, Ph.D.
Professor of Medical Genetics
Mercer University School of Medicine
Macon, Georgia

Harold K. Voris, Ph.D.
Curator and Head
Division of Amphibians and Reptiles
The Field Museum of Natural History
Chicago, Illinois

Mollie Walton
Biology Instructor
El Paso Community College
El Paso, Texas

Peter Wetherwax, Ph.D.
Professor of Biology
University of Oregon
Eugene, Oregon

Mary K. Wicksten, Ph.D.
Professor of Biology
Texas A&M University
College Station, Texas

R. Stimson Wilcox, Ph.D.
Associate Professor of Biology
Department of Biological Sciences
Binghamton University
Binghamton, New York

Conrad M. Zapanta, Ph.D.
Research Engineer
Sulzer Carbomedics, Inc.
Austin, Texas

Safety Reviewer

Jack Gerlovich, Ph.D.
Associate Professor
School of Education
Drake University
Des Moines, Iowa

Teacher Reviewers

Barry L. Bishop
Science Teacher and Dept. Chair
San Rafael Junior High School
Ferron, Utah

Carol A. Bornhorst
Science Teacher and Dept. Chair
Bonita Vista Middle School
Chula Vista, California

Paul Boyle
Science Teacher
Perry Heights Middle School
Evansville, Indiana

Yvonne Brannum
Science Teacher and Dept. Chair
Hine Junior High School
Washington, D.C.

Gladys Cherniak
Science Teacher
St. Paul's Episcopal School
Mobile, Alabama

James Chin
Science Teacher
Frank A. Day Middle School
Newtonville, Massachusetts

Kenneth Creese
Science Teacher
White Mountain Junior High School
Rock Springs, Wyoming

Linda A. Culp
Science Teacher and Dept. Chair
Thorndale High School
Thorndale, Texas

Georgiann Delgadillo
Science Teacher
East Valley Continuous Curriculum School
Spokane, Washington

Alonda Droege
Biology Teacher
Evergreen High School
Seattle, Washington

Michael J. DuPré
Curriculum Specialist
Rush Henrietta Junior-Senior High School
Henrietta, New York

Rebecca Ferguson
Science Teacher
North Ridge Middle School
North Richland Hills, Texas

Susan Gorman
Science Teacher
North Ridge Middle School
North Richland Hills, Texas

Gary Habeeb
Science Mentor
Sierra-Plumas Joint Unified School District
Downieville, California

Karma Houston-Hughes
Science Mentor
Kyrene Middle School
Tempe, Arizona

Roberta Jacobowitz
Science Teacher
C. W. Otto Middle School
Lansing, Michigan

Kerry A. Johnson
Science Teacher
Isbell Middle School
Santa Paula, California

M. R. Penny Kisiah
Science Teacher and Dept. Chair
Fairview Middle School
Tallahassee, Florida

Kathy LaRoe
Science Teacher
East Valley Middle School
East Helena, Montana

Jane M. Lemons
Science Teacher
Western Rockingham Middle School
Madison, North Carolina

Scott Mandel, Ph.D.
Director and Educational Consultant
Teachers Helping Teachers
Los Angeles, California

Thomas Manerchia
Former Biology and Life Science Teacher
Archmere Academy
Claymont, Delaware

Maurine O. Marchani
Science Teacher and Dept. Chair
Raymond Park Middle School
Indianapolis, Indiana

Jason P. Marsh
Biology Teacher
Montevideo High School and Montevideo Country School
Montevideo, Minnesota

Edith C. McAlanis
Science Teacher and Dept. Chair
Socorro Middle School
El Paso, Texas

Kevin McCurdy, Ph.D.
Science Teacher
Elmwood Junior High School
Rogers, Arkansas

Kathy McKee
Science Teacher
Hoyt Middle School
Des Moines, Iowa

Acknowledgments continue on page 211.

A Microorganisms, Fungi, and Plants

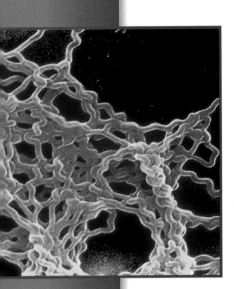

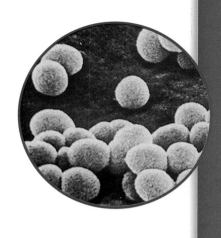

Skills Development

Process Skills

QuickLabs

Chapter Labs

Skills Development *(continued)*

Research and Critical Thinking Skills

Apply

Feature Articles

Connections

Mathematics

Program Scope and Sequence

Selecting the right books for your course is easy. Just review the topics presented in each book to determine the best match to your district curriculum.

	A MICROORGANISMS, FUNGI, AND PLANTS	**B** ANIMALS	
CHAPTER 1	**It's Alive!! Or, Is It?** ❏ Characteristics of living things ❏ Homeostasis ❏ Heredity and DNA ❏ Producers, consumers, and decomposers ❏ Biomolecules	**Animals and Behavior** ❏ Characteristics of animals ❏ Classification of animals ❏ Animal behavior ❏ Hibernation and estivation ❏ The biological clock ❏ Animal communication ❏ Living in groups	
CHAPTER 2	**Bacteria and Viruses** ❏ Binary fission ❏ Characteristics of bacteria ❏ Nitrogen-fixing bacteria ❏ Antibiotics ❏ Pathogenic bacteria ❏ Characteristics of viruses ❏ Lytic cycle	**Invertebrates** ❏ General characteristics of invertebrates ❏ Types of symmetry ❏ Characteristics of sponges, cnidarians, arthropods, and echinoderms ❏ Flatworms versus roundworms ❏ Types of circulatory systems	
CHAPTER 3	**Protists and Fungi** ❏ Characteristics of protists ❏ Types of algae ❏ Types of protozoa ❏ Protist reproduction ❏ Characteristics of fungi and lichens	**Fishes, Amphibians, and Reptiles** ❏ Characteristics of vertebrates ❏ Structure and kinds of fishes ❏ Development of lungs ❏ Structure and kinds of amphibians and reptiles ❏ Function of the amniotic egg	
CHAPTER 4	**Introduction to Plants** ❏ Characteristics of plants and seeds ❏ Reproduction and classification ❏ Angiosperms versus gymnosperms ❏ Monocots versus dicots ❏ Structure and functions of roots, stems, leaves, and flowers	**Birds and Mammals** ❏ Structure and kinds of birds ❏ Types of feathers ❏ Adaptations for flight ❏ Structure and kinds of mammals ❏ Function of the placenta	
CHAPTER 5	**Plant Processes** ❏ Pollination and fertilization ❏ Dormancy ❏ Photosynthesis ❏ Plant tropisms ❏ Seasonal responses of plants		
CHAPTER 6			
CHAPTER 7			

Life Science

C — CELLS, HEREDITY, & CLASSIFICATION

Cells: The Basic Units of Life
- ❏ Cells, tissues, and organs
- ❏ Populations, communities, and ecosystems
- ❏ Cell theory
- ❏ Surface-to-volume ratio
- ❏ Prokaryotic versus eukaryotic cells
- ❏ Cell organelles

The Cell in Action
- ❏ Diffusion and osmosis
- ❏ Passive versus active transport
- ❏ Endocytosis versus exocytosis
- ❏ Photosynthesis
- ❏ Cellular respiration and fermentation
- ❏ Cell cycle

Heredity
- ❏ Dominant versus recessive traits
- ❏ Genes and alleles
- ❏ Genotype, phenotype, the Punnett square and probability
- ❏ Meiosis
- ❏ Determination of sex

Genes and Gene Technology
- ❏ Structure of DNA
- ❏ Protein synthesis
- ❏ Mutations
- ❏ Heredity disorders and genetic counseling

The Evolution of Living Things
- ❏ Adaptations and species
- ❏ Evidence for evolution
- ❏ Darwin's work and natural selection
- ❏ Formation of new species

The History of Life on Earth
- ❏ Geologic time scale and extinctions
- ❏ Plate tectonics
- ❏ Human evolution

Classification
- ❏ Levels of classification
- ❏ Cladistic diagrams
- ❏ Dichotomous keys
- ❏ Characteristics of the six kingdoms

D — HUMAN BODY SYSTEMS & HEALTH

Body Organization and Structure
- ❏ Homeostasis
- ❏ Types of tissue
- ❏ Organ systems
- ❏ Structure and function of the skeletal system, muscular system, and integumentary system

Circulation and Respiration
- ❏ Structure and function of the cardiovascular system, lymphatic system, and respiratory system
- ❏ Respiratory disorders

The Digestive and Urinary Systems
- ❏ Structure and function of the digestive system
- ❏ Structure and function of the urinary system

Communication and Control
- ❏ Structure and function of the nervous system and endocrine system
- ❏ The senses
- ❏ Structure and function of the eye and ear

Reproduction and Development
- ❏ Asexual versus sexual reproduction
- ❏ Internal versus external fertilization
- ❏ Structure and function of the human male and female reproductive systems
- ❏ Fertilization, placental development, and embryo growth
- ❏ Stages of human life

Body Defenses and Disease
- ❏ Types of diseases
- ❏ Vaccines and immunity
- ❏ Structure and function of the immune system
- ❏ Autoimmune diseases, cancer, and AIDS

Staying Healthy
- ❏ Nutrition and reading food labels
- ❏ Alcohol and drug effects on the body
- ❏ Hygiene, exercise, and first aid

E — ENVIRONMENTAL SCIENCE

Interactions of Living Things
- ❏ Biotic versus abiotic parts of the environment
- ❏ Producers, consumers, and decomposers
- ❏ Food chains and food webs
- ❏ Factors limiting population growth
- ❏ Predator-prey relationships
- ❏ Symbiosis and coevolution

Cycles in Nature
- ❏ Water cycle
- ❏ Carbon cycle
- ❏ Nitrogen cycle
- ❏ Ecological succession

The Earth's Ecosystems
- ❏ Kinds of land and water biomes
- ❏ Marine ecosystems
- ❏ Freshwater ecosystems

Environmental Problems and Solutions
- ❏ Types of pollutants
- ❏ Types of resources
- ❏ Conservation practices
- ❏ Species protection

Energy Resources
- ❏ Types of resources
- ❏ Energy resources and pollution
- ❏ Alternative energy resources

Scope and Sequence *(continued)*

	F INSIDE THE RESTLESS EARTH	**G** EARTH'S CHANGING SURFACE
CHAPTER 1	**Minerals of the Earth's Crust** ❏ Mineral composition and structure ❏ Types of minerals ❏ Mineral identification ❏ Mineral formation and mining	**Maps as Models of the Earth** ❏ Structure of a map ❏ Cardinal directions ❏ Latitude, longitude, and the equator ❏ Magnetic declination and true north ❏ Types of projections ❏ Aerial photographs ❏ Remote sensing ❏ Topographic maps
CHAPTER 2	**Rocks: Mineral Mixtures** ❏ Rock cycle and types of rocks ❏ Rock classification ❏ Characteristics of igneous, sedimentary, and metamorphic rocks	**Weathering and Soil Formation** ❏ Types of weathering ❏ Factors affecting the rate of weathering ❏ Composition of soil ❏ Soil conservation and erosion prevention
CHAPTER 3	**The Rock and Fossil Record** ❏ Uniformitarianism versus catastrophism ❏ Superposition ❏ The geologic column and unconformities ❏ Absolute dating and radiometric dating ❏ Characteristics and types of fossils ❏ Geologic time scale	**Agents of Erosion and Deposition** ❏ Shoreline erosion and deposition ❏ Wind erosion and deposition ❏ Erosion and deposition by ice ❏ Gravity's effect on erosion and deposition
CHAPTER 4	**Plate Tectonics** ❏ Structure of the Earth ❏ Continental drifts and sea floor spreading ❏ Plate tectonics theory ❏ Types of boundaries ❏ Types of crust deformities	
CHAPTER 5	**Earthquakes** ❏ Seismology ❏ Features of earthquakes ❏ P and S waves ❏ Gap hypothesis ❏ Earthquake safety	
CHAPTER 6	**Volcanoes** ❏ Types of volcanoes and eruptions ❏ Types of lava and pyroclastic material ❏ Craters versus calderas ❏ Sites and conditions for volcano formation ❏ Predicting eruptions	

Earth Science

Scope and Sequence (continued)

	K INTRODUCTION TO MATTER	**L** INTERACTIONS OF MATTER
CHAPTER 1	**The Properties of Matter** ❏ Definition of matter ❏ Mass and weight ❏ Physical and chemical properties ❏ Physical and chemical change ❏ Density	**Chemical Bonding** ❏ Types of chemical bonds ❏ Valence electrons ❏ Ions versus molecules ❏ Crystal lattice
CHAPTER 2	**States of Matter** ❏ States of matter and their properties ❏ Boyle's and Charles's laws ❏ Changes of state	**Chemical Reactions** ❏ Writing chemical formulas and equations ❏ Law of conservation of mass ❏ Types of reactions ❏ Endothermic versus exothermic reactions ❏ Law of conservation of energy ❏ Activation energy ❏ Catalysts and inhibitors
CHAPTER 3	**Elements, Compounds, and Mixtures** ❏ Elements and compounds ❏ Metals, nonmetals, and metalloids (semiconductors) ❏ Properties of mixtures ❏ Properties of solutions, suspensions, and colloids	**Chemical Compounds** ❏ Ionic versus covalent compounds ❏ Acids, bases, and salts ❏ pH ❏ Organic compounds ❏ Biomolecules
CHAPTER 4	**Introduction to Atoms** ❏ Atomic theory ❏ Atomic model and structure ❏ Isotopes ❏ Atomic mass and mass number	**Atomic Energy** ❏ Properties of radioactive substances ❏ Types of decay ❏ Half-life ❏ Fission, fusion, and chain reactions
CHAPTER 5	**The Periodic Table** ❏ Structure of the periodic table ❏ Periodic law ❏ Properties of alkali metals, alkaline-earth metals, halogens, and noble gases	
CHAPTER 6		

Physical Science

M — FORCES, MOTION, AND ENERGY

Matter in Motion
- ❏ Speed, velocity, and acceleration
- ❏ Measuring force
- ❏ Friction
- ❏ Mass versus weight

Forces in Motion
- ❏ Terminal velocity and free fall
- ❏ Projectile motion
- ❏ Inertia
- ❏ Momentum

Forces in Fluids
- ❏ Properties in fluids
- ❏ Atmospheric pressure
- ❏ Density
- ❏ Pascal's principle
- ❏ Buoyant force
- ❏ Archimedes' principle
- ❏ Bernoulli's principle

Work and Machines
- ❏ Measuring work
- ❏ Measuring power
- ❏ Types of machines
- ❏ Mechanical advantage
- ❏ Mechanical efficiency

Energy and Energy Resources
- ❏ Forms of energy
- ❏ Energy conversions
- ❏ Law of conservation of energy
- ❏ Energy resources

Heat and Heat Technology
- ❏ Heat versus temperature
- ❏ Thermal expansion
- ❏ Absolute zero
- ❏ Conduction, convection, radiation
- ❏ Conductors versus insulators
- ❏ Specific heat capacity
- ❏ Changes of state
- ❏ Heat engines
- ❏ Thermal pollution

N — ELECTRICITY AND MAGNETISM

Introduction to Electricity
- ❏ Law of electric charges
- ❏ Conduction versus induction
- ❏ Static electricity
- ❏ Potential difference
- ❏ Cells, batteries, and photocells
- ❏ Thermocouples
- ❏ Voltage, current, and resistance
- ❏ Electric power
- ❏ Types of circuits

Electromagnetism
- ❏ Properties of magnets
- ❏ Magnetic force
- ❏ Electromagnetism
- ❏ Solenoids and electric motors
- ❏ Electromagnetic induction
- ❏ Generators and transformers

Electronic Technology
- ❏ Properties of semiconductors
- ❏ Integrated circuits
- ❏ Diodes and transistors
- ❏ Analog versus digital signals
- ❏ Microprocessors
- ❏ Features of computers

O — SOUND AND LIGHT

The Energy of Waves
- ❏ Properties of waves
- ❏ Types of waves
- ❏ Reflection and refraction
- ❏ Diffraction and interference
- ❏ Standing waves and resonance

The Nature of Sound
- ❏ Properties of sound waves
- ❏ Structure of the human ear
- ❏ Pitch and the Doppler effect
- ❏ Infrasonic versus ultrasonic sound
- ❏ Sound reflection and echolocation
- ❏ Sound barrier
- ❏ Interference, resonance, diffraction, and standing waves
- ❏ Sound quality of instruments

The Nature of Light
- ❏ Electromagnetic waves
- ❏ Electromagnetic spectrum
- ❏ Law of reflection
- ❏ Absorption and scattering
- ❏ Reflection and refraction
- ❏ Diffraction and interference

Light and Our World
- ❏ Luminosity
- ❏ Types of lighting
- ❏ Types of mirrors and lenses
- ❏ Focal point
- ❏ Structure of the human eye
- ❏ Lasers and holograms

Components Listing

Effective planning starts with all the resources you need in an easy-to-use package for each short course.

Directed Reading Worksheets Help students develop and practice fundamental reading comprehension skills and provide a comprehensive review tool for students to use when studying for an exam.

Study Guide Vocabulary & Notes Worksheets and Chapter Review Worksheets are reproductions of the Chapter Highlights and Chapter Review sections that follow each chapter in the textbook.

Science Puzzlers, Twisters & Teasers Use vocabulary and concepts from each chapter of the Pupil's Editions as elements of rebuses, anagrams, logic puzzles, daffy definitions, riddle poems, word jumbles, and other types of puzzles.

Reinforcement and Vocabulary Review Worksheets Approach a chapter topic from a different angle with an emphasis on different learning modalities to help students that are frustrated by traditional methods.

Critical Thinking & Problem Solving Worksheets Develop the following skills: distinguishing fact from opinion, predicting consequences, analyzing information, and drawing conclusions. Problem Solving Worksheets develop a step-by-step process of problem analysis including gathering information, asking critical questions, identifying alternatives, and making comparisons.

Math Skills for Science Worksheets Each activity gives a brief introduction to a relevant math skill, a step-by-step explanation of the math process, one or more example problems, and a variety of practice problems.

Science Skills Worksheets Help your students focus specifically on skills such as measuring, graphing, using logic, understanding statistics, organizing research papers, and critical thinking options.

LAB ACTIVITIES

Datasheets for Labs These worksheets are the labs found in the *Holt Science & Technology* textbook. Charts, tables, and graphs are included to make data collection and analysis easier, and space is provided to write observations and conclusions.

Whiz-Bang Demonstrations Discovery or Making Models experiences label each demo as one in which students discover an answer or use a scientific model.

Calculator-Based Labs Give students the opportunity to use graphing-calculator probes and sensors to collect data using a TI graphing calculator, Vernier sensors, and a TI CBL 2™ or Vernier Lab Pro interface.

EcoLabs and Field Activities Focus on educational outdoor projects, such as wildlife observation, nature surveys, or natural history.

Inquiry Labs Use the scientific method to help students find their own path in solving a real-world problem.

Long-Term Projects and Research Ideas Provide students with the opportunity to go beyond library and Internet resources to explore science topics.

ASSESSMENT

Chapter Tests Each four-page chapter test consists of a variety of item types including Multiple Choice, Using Vocabulary, Short Answer, Critical Thinking, Math in Science, Interpreting Graphics, and Concept Mapping.

Performance-Based Assessments Evaluate students' abilities to solve problems using the tools, equipment, and techniques of science. Rubrics included for each assessment make it easy to evaluate student performance.

TEACHER RESOURCES

Lesson Plans Integrate all of the great resources in the *Holt Science & Technology* program into your daily teaching. Each lesson plan includes a correlation of the lesson activities to the National Science Education Standards.

Teaching Transparencies Each transparency is correlated to a particular lesson in the Chapter Organizer.

ALL LABS ARE CLASSROOM TESTED & APPROVED

Concept Mapping Transparencies, Worksheets, and Answer Key

Give students an opportunity to complete their own concept maps to study the concepts within each chapter and form logical connections. Student worksheets contain a blank concept map with linking phrases and a list of terms to be used by the student to complete the map.

TECHNOLOGY RESOURCES

 One-Stop Planner CD-ROM

Finding the right resources is easy with the One-Stop Planner CD-ROM. You can view and print any resource with just the click of a mouse. Customize the suggested lesson plans to match your daily or weekly calendar and your district's requirements. Powerful test generator software allows you to create customized assessments using a databank of items.

The One-Stop Planner for each level includes the following:

- All materials from the Teaching Resources
- Bellringer Transparency Masters
- Block Scheduling Tools
- Standards Correlations
- Lab Inventory Checklist
- Safety Information
- Science Fair Guide
- Parent Involvement Tools
- Spanish Audio Scripts
- Spanish Glossary
- Assessment Item Listing
- Assessment Checklists and Rubrics
- Test Generator

sciLINKS

*sci*LINKS numbers throughout the text take you and your students to some of the best on-line resources available. Sites are constantly reviewed and updated by the National Science Teachers Association. Special "teacher only" sites are available to you once you register with the service.

 go.hrw.com

To access Holt, Rinehart and Winston Web resources, use the home page codes for each level found on page 1 of the Pupil's Editions. The codes shown on the Chapter Organizers for each chapter in the Annotated Teacher's Edition take you to chapter-specific resources.

 Smithsonian Institution

Find lesson plans, activities, interviews, virtual exhibits, and just general information on a wide variety of topics relevant to middle school science.

CNNfyi.com

Find the latest in late-breaking science news for students. Featured news stories are supported with lesson plans and activities.

 Presents Science in the News Video Library

Bring relevant science news stories into the classroom. Each video comes with a Teacher's Guide and set of Critical Thinking Worksheets that develop listening and media analysis skills. Tapes in the series include:

- Eye on the Environment
- Multicultural Connections
- Scientists in Action
- Science, Technology & Society

 Guided Reading Audio CD Program

Students can listen to a direct read of each chapter and follow along in the text. Use the program as a content bridge for struggling readers and students for whom English is not their native language.

 Interactive Explorations CD-ROM

Turn a computer into a virtual laboratory. Students act as lab assistants helping Dr. Crystal Labcoat solve real-world problems. Activities develop students' inquiry, analysis, and decision-making skills.

 Interactive Science Encyclopedia CD-ROM

Give your students access to more than 3,000 cross-referenced scientific definitions, in-depth articles, science fair project ideas, activities, and more.

ADDITIONAL COMPONENTS

Holt Anthology of Science Fiction

Science Fiction features in the Pupil's Edition preview the stories found in the anthology. Each story begins with a Reading Prep guide and closes with Think About It questions.

Professional Reference for Teachers

Articles written by leading educators help you learn more about the National Science Education Standards, block scheduling, classroom management techniques, and more. A bibliography of professional references is included.

Holt Science Posters

Seven wall posters highlight interesting topics, such as the Physics of Sports, or useful reference material, such as the Scientific Method.

 Holt Science Skills Workshop: Reading in the Content Area

Use a variety of in-depth skills exercises to help students learn to read science materials strategically.

 Key

These materials are blackline masters.

All titles shown in green are found in the *Teaching Resources* booklets for each course.

Science & Math Skills Worksheets

The *Holt Science and Technology* program helps you meet the needs of a wide variety of students, regardless of their skill level. The following pages provide examples of the worksheets available to improve your students' science and math skills, whether they already have a strong science and math background or are weak in these areas. Samples of assessment checklists and rubrics are also provided.

In addition to the skills worksheets represented here, *Holt Science and Technology* provides a variety of worksheets that are correlated directly with each chapter of the program. Representations of these worksheets are found at the beginning of each chapter in this Annotated Teacher's Edition. Specific worksheets related to each chapter are listed in the Chapter Organizer. Worksheets and transparencies are found in the softcover *Teaching Resources* for each course.

Many worksheets are also available on the HRW Web site. The address is **go.hrw.com.**

Science Skills Worksheets: Thinking Skills

BEING FLEXIBLE
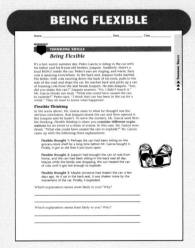

USING YOUR SENSES

THINKING OBJECTIVELY
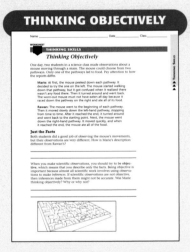

UNDERSTANDING BIAS
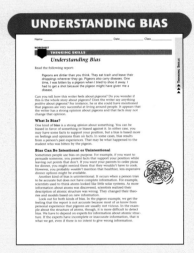

USING LOGIC

BOOSTING YOUR MEMORY
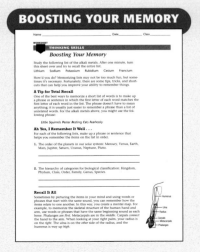

IMPROVING YOUR STUDY HABITS

READING A SCIENCE TEXTBOOK
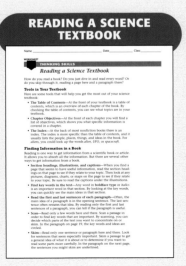

Science Skills Worksheets: Experimenting Skills

SAFETY RULES!
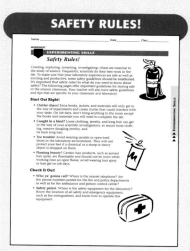

DOING A LAB WRITE-UP
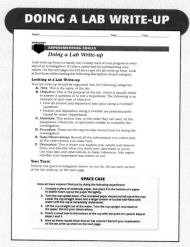

UNDERSTANDING VARIABLES
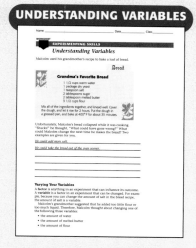

WORKING WITH HYPOTHESES
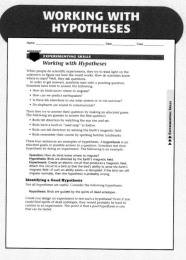

DESIGNING AN EXPERIMENT
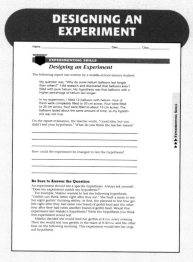

USING THE INTERNATIONAL SYSTEM OF UNITS (SI)
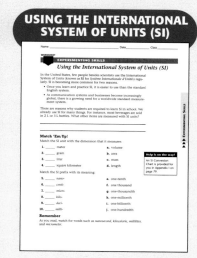

MEASURING
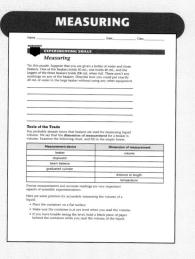

Science Skills Worksheets: Researching Skills

CHOOSING YOUR TOPIC
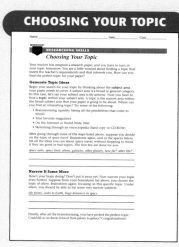

ORGANIZING YOUR RESEARCH
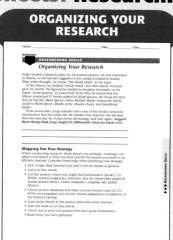

FINDING USEFUL SOURCES
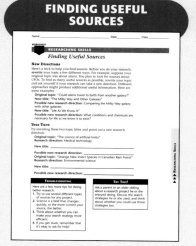

RESEARCHING ON THE WEB
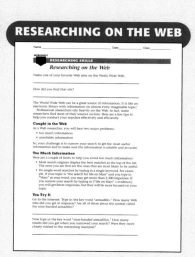

Science & Math Skills Worksheets (continued)

Science Skills Worksheets: Researching Skills (continued)

IDENTIFYING BIAS

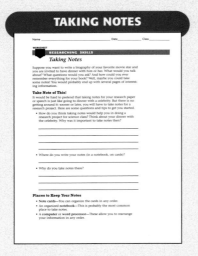

TAKING NOTES

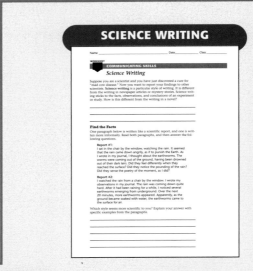

SCIENCE WRITING

Science Skills Worksheets: Communicating Skills

SCIENCE DRAWING

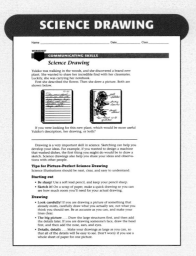

USING MODELS TO COMMUNICATE

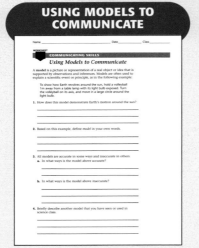

INTRODUCTION TO GRAPHS

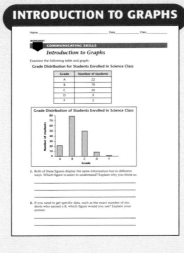

GRASPING GRAPHING

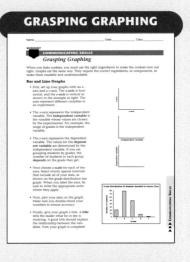

INTERPRETING YOUR DATA

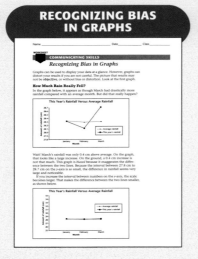

RECOGNIZING BIAS IN GRAPHS

MAKING DATA MEANINGFUL

HINTS FOR ORAL PRESENTATIONS

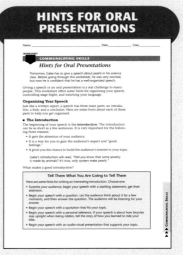

Math Skills for Science

ADDITION AND SUBTRACTION
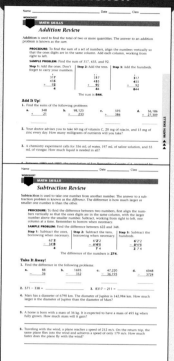

MULTIPLICATION
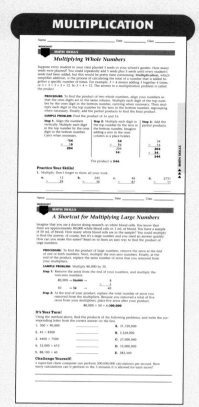

DIVISION

AVERAGES
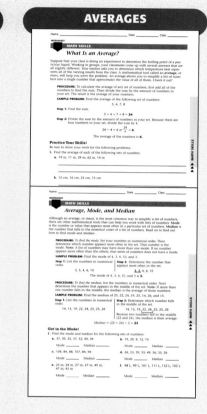

POSITIVE AND NEGATIVE NUMBERS
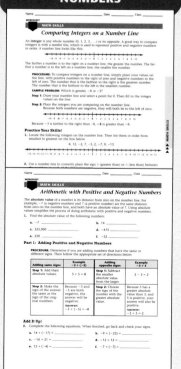

FRACTIONS
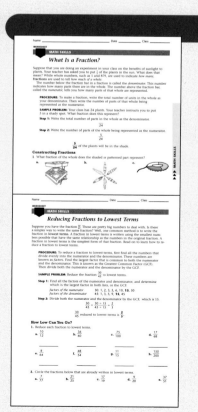

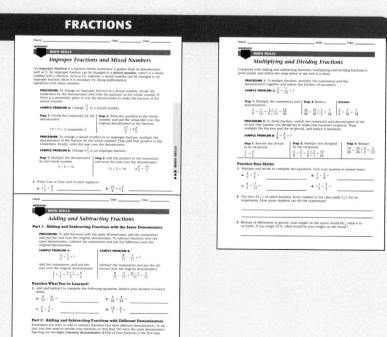

Science & Math Skills Worksheets

Math Skills for Science (continued)

RATIOS AND PROPORTIONS

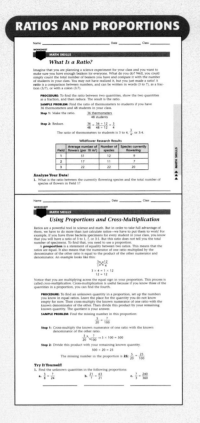

DECIMALS

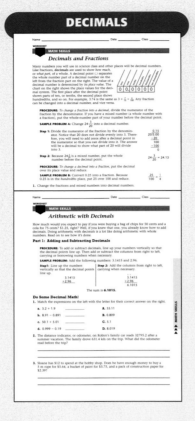

PERCENTAGES

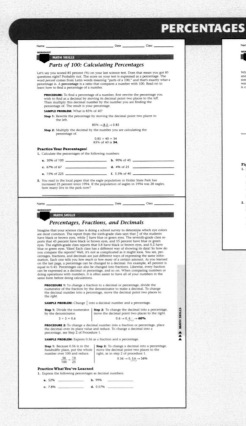

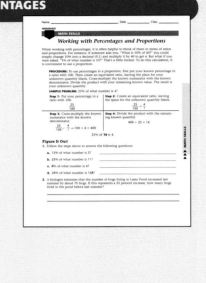

POWERS OF 10

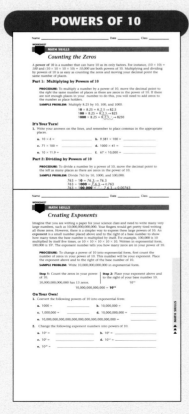

SCIENTIFIC NOTATION

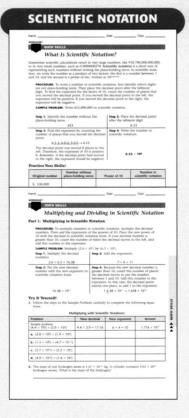

SI MEASUREMENT AND CONVERSION

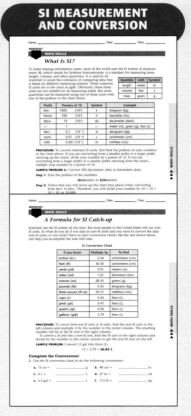

Math Skills for Science (continued)

GEOMETRY

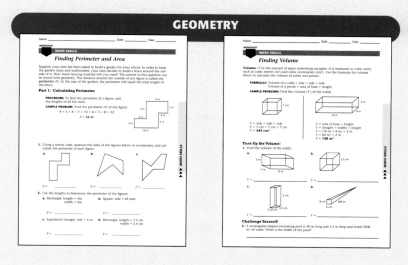

Finding Perimeter and Area

Suppose your class has been asked to build a garden for your school. In order to keep the garden clean and undisturbed, your class decides to build a fence around the outside of it. How much fencing material will you need? The answer to this question can be found with geometry. The distance around the outside of any figure is called the *perimeter* (*P*). In the case of the garden, the perimeter will equal the total length of the fence.

Part 1: Calculating Perimeter

PROCEDURE: To find the perimeter of a figure, add the lengths of all the sides.

SAMPLE PROBLEM: Find the perimeter (*P*) of the figure.
$9 + 5 + 4 + 7 + 10 + 4 + 5 + 8 = 52$
$P = 52\ \text{m}$

1. Using a metric rule, measure the sides of the figures below in centimeters, and calculate the perimeter of each figure.

a.   b.   c.

$P =$ $P =$ $P =$

2. Use the lengths to determine the perimeter of the figures.
a. Rectangle: length = 4 m, width = 2m
b. Square: side = 45 mm
c. Equilateral triangle: side = 6 m
d. Rectangle: length = 3.5 cm, width = 2.4 cm

Finding Volume

Volume (*V*) is the amount of space something occupies. It is measured in cubic units, such as cubic meters (m³) and cubic centimeters (cm³). Use the formulas for volume below to calculate the volume of cubes and prisms.

FORMULAS: Volume of a cube = side × side × side
Volume of a prism = area of base × height

SAMPLE PROBLEMS: Find the volume (*V*) of the solids.

$V = \text{side} \times \text{side} \times \text{side}$
$V = 7\ \text{cm} \times 7\ \text{cm} \times 7\ \text{cm}$
$V = 343\ \text{cm}^3$

$V = \text{area of base} \times \text{height}$
$V = (\text{length} \times \text{width}) \times \text{height}$
$V = (16\ \text{m} \times 4\ \text{m}) \times 2\ \text{m}$
$V = 64\ \text{m}^2 \times 2\ \text{m}$
$V = 128\ \text{m}^3$

Turn Up the Volume!

1. Find the volume of the solids.

a.   b.   c.   d.

$V =$

Challenge Yourself!

2. A rectangular-shaped swimming pool is 50 m long and 2.5 m deep and holds 2500 m³ of water. What is the width of the pool?

THE UNIT FACTOR AND DIMENSIONAL ANALYSIS

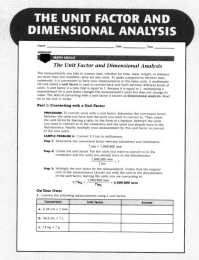

The Unit Factor and Dimensional Analysis

The measurements you take in science class, whether for time, mass, weight, or distance, are more than just numbers—they are also units. To make comparisons between measurements, it is convenient to have your measurements in the same units. A mathematical tool called a **unit factor** is used to convert back and forth between different kinds of units. A unit factor is a ratio that is equal to 1. Because it is equal to 1, multiplying a measurement by a unit factor changes the measurement's units but does not change its value. The skill of converting with a unit factor is known as **dimensional analysis**. Read on to see how it works.

Part 1: Converting with a Unit Factor

PROCEDURE: To convert units with a unit factor, determine the conversion factor between the units you have and the units you want to convert to. Then create the unit factor by making a ratio, in the form of a fraction, between the units you want to convert to in the numerator and the units you already have in the denominator. Finally, multiply your measurement by this unit factor to convert to the new units.

SAMPLE PROBLEM A: Convert 3.5 km to millimeters.

Step 1: Determine the conversion factor between kilometers and millimeters.
$1\ \text{km} = 1,000,000\ \text{mm}$

Step 2: Create the unit factor. Put the units you want to convert to in the numerator and the units you already have in the denominator.
$\dfrac{1,000,000\ \text{mm}}{1\ \text{km}} = 1$

Step 3: Multiply the unit factor by the measurement. Notice that the original unit of the measurement cancels out with the unit in the denominator of the unit factor, leaving the units you are converting to.
$3.5\ \text{km} \times \dfrac{1,000,000\ \text{mm}}{1\ \text{km}} = 3,500,000\ \text{mm}$

On Your Own!

1. Convert the following measurements using a unit factor:

Conversion	Unit factor	Answer
a. 2.34 cm = ? mm		
b. 54.6 mL = ? L		
c. 12 kg = ? g		

MATH IN SCIENCE: INTEGRATED SCIENCE

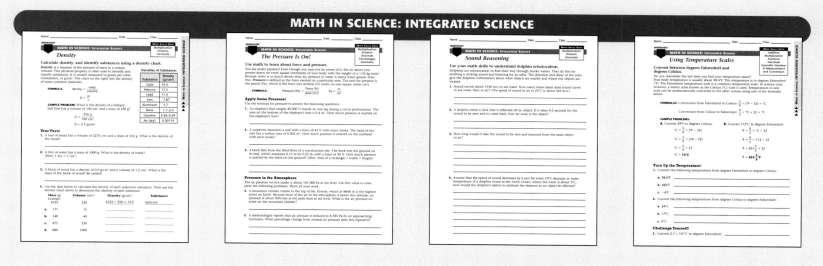

Density

Calculate density, and identify substances using a density chart.

Density is a measure of the amount of mass in a certain volume. This physical property is often used to identify and classify substances. It is usually measured in grams per cubic centimeters, or g/cm³. The chart on the right lists the density of some common materials.

FORMULA: $\text{density} = \dfrac{\text{mass}}{\text{volume}}$
$D = \dfrac{m}{v}$

SAMPLE PROBLEM: What is the density of a billiard ball that has a volume of 100 cm³ and a mass of 250 g?
$D = \dfrac{250\ \text{g}}{100\ \text{cm}^3}$
$D = 2.5\ \text{g/cm}^3$

Densities of Substances

Substance	Density (g/cm³)
Gold	19.3
Mercury	13.5
Lead	11.4
Iron	7.87
Aluminum	3.7
Bone	1.7–2.0
Gasoline	0.66–0.69
Air (dry)	0.00119

Your Turn!

1. A loaf of bread has a volume of 2270 cm³ and a mass of 454 g. What is the density of the bread?

2. A liter of water has a mass of 1000 g. What is the density of water? (Hint: 1 mL = 1 cm³)

3. A block of wood has a density of 0.6 g/cm³ and a volume of 1.2 cm³. What is the mass of the block of wood? Be careful!

4. Use the data below to calculate the density of each unknown substance. Then use the density chart above to determine the identity of each substance.

Mass (g)	Volume (cm³)	Density (g/cm³)	Substance
Example: 4725	350	4725 ÷ 350 = 13.5	mercury
a. 171	15		
b. 148	40		
c. 473	250		
d. 680	1000		

Math Skills Used: Multiplication, Division, Decimals

The Pressure Is On!

Use math to learn about force and pressure.

You are under pressure! Even though you may not be aware of it, the air above you presses down on every square centimeter of your body with the weight of a 1.03 kg mass! Because water is so much denser than air, pressure in water is many times greater than this. **Pressure** is the force exerted on a particular area. The unit for pressure is the pascal (Pa), which is the force one newton (N) exerts on one square meter (m²).

Apply Some Pressure!

Use the formula for pressure to answer the following questions:

$\text{Pressure (Pa)} = \dfrac{\text{Force (N)}}{\text{Area (m}^2)} \qquad Pa = \dfrac{N}{m^2}$

1. An elephant that weighs 40,000 N stands on one leg during a circus performance. The area on the bottom of the elephant's foot is 0.4 m². How much pressure is exerted on the elephant's foot?

2. A carpenter hammers a nail with a force of 45 N with every stroke. The head of the nail has a surface area of 0.002 m². How much pressure is exerted on the nailhead with each stroke?

3. A brick falls from the third floor of a construction site. The brick hits the ground on its end, which measures 0.15 m by 0.25 m, with a force of 30 N. How much pressure is exerted by the brick on the ground? (Hint: Area of a rectangle = width × length)

Pressure in the Atmosphere

The air pressure we live under is about 101,000 Pa at sea level. Use this value to complete the following problems. Show all your work.

4. A mountain climber climbs to the top of Mt. Everest, which at 8848 m is the highest point on Earth. Because most of the air in the atmosphere is below this altitude, air pressure is about 50% less at the peak than at sea level. What is the air pressure exerted on the mountain climber?

5. A meteorologist reports that air pressure is reduced to 8,585 Pa by an approaching hurricane. What percentage change from normal air pressure does this represent?

Math Skills Used: Multiplication, Division, Decimals, Percentages, Geometry

Sound Reasoning

Use your math skills to understand dolphin echolocation.

Dolphins use echolocation to find their way through murky waters. They do this by emitting a clicking sound and listening for an echo. The direction and delay of the echo give the dolphins information about what objects are nearby and where the objects are located.

1. Sound travels about 1530 m/s in sea water. How many times faster does sound travel in sea water than in air? (The speed of sound in air at 25°C is about 345 m/s.)

2. A dolphin emits a click that is reflected off an object. If it takes 0.2 seconds for the sound to be sent and to come back, how far away is the object?

3. How long would it take the sound to be sent and returned from the same object in air?

4. Assume that the speed of sound decreases by 6 m/s for every 10°C decrease in water temperature. If a dolphin swam to the Arctic Ocean, where the water is about 5°C, how would the dolphin's ability to estimate the distance to an object be affected?

Math Skills Used: Multiplication, Division, Decimals

Using Temperature Scales

Convert between degrees Fahrenheit and degrees Celsius.

Do you remember the last time you had your temperature taken? Your body temperature is usually about 98.6°F. This temperature is in degrees Fahrenheit (°F). The Fahrenheit temperature scale is a common temperature scale. In science class, however, a metric scale known as the Celsius (°C) scale is used. Temperatures in one scale can be mathematically converted to the other system using one of the formulas below.

FORMULAS: Conversion from Fahrenheit to Celsius: $\frac{5}{9} \times (°F - 32) = °C$
Conversion from Celsius to Fahrenheit: $\frac{9}{5} \times °C + 32 = °F$

SAMPLE PROBLEMS:

A. Convert 59°F to degrees Celsius.
$°C = \frac{5}{9} \times (°F - 32)$
$°C = \frac{5}{9} \times (59 - 32)$
$°C = \frac{5}{9} \times 27$
$°C = 15°C$

B. Convert 112°C to degrees Fahrenheit.
$°F = \frac{9}{5} \times °C + 32$
$°F = \frac{9}{5} \times 112 + 32$
$°F = 201\frac{3}{5} + 32$
$°F = 233\frac{3}{5}°F$

Turn Up the Temperature!

1. Convert the following temperatures from degrees Fahrenheit to degrees Celsius:
a. 98.6°F
b. 482°F
c. −4°F

2. Convert the following temperatures from degrees Celsius to degrees Fahrenheit:
a. 24°C
b. 17°C
c. 0°C

Challenge Yourself!

3. Convert 2.7 × 10⁴°C to degrees Fahrenheit.

Math Skills Used: Addition, Multiplication, Fractions, Decimals, Scientific Notation, SI Measurement and Conversion

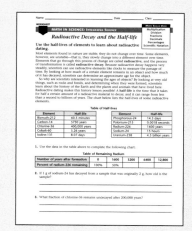

Radioactive Decay and the Half-life

Use the half-lives of elements to learn about radioactive dating.

Most elements found in nature are stable; they do not change over time. Some elements, however, are unstable—that is, they slowly change into a different element over time. Elements that go through this process of change are called **radioactive**, and the process of transformation is called **radioactive decay**. Because radioactive decay happens very steadily, scientists can use radioactive elements like clocks to measure the passage of time. By looking at how much of a certain element remains in an object and how much of it has decayed, scientists can determine an approximate age for the object.

So why are scientists interested in learning the ages of objects? By looking at very old things, such as rocks and fossils, and determining when they were formed, scientists learn about the history of the Earth and the plants and animals that have lived here. Radioactive dating makes this history lesson possible! A **half-life** is the time that it takes for half a certain amount of a radioactive material to decay, and it can range from less than a second to billions of years. The chart below lists the half-lives of some radioactive elements.

Table of Half-lives

Element	Half-life	Element	Half-life
Bismuth-212	60.5 minutes	Phosphorous-24	14.3 days
Carbon-14	5730 years	Polonium-215	0.0018 seconds
Chlorine-36	400,000 years	Radium-226	1600 years
Cobalt-60	5.26 years	Sodium-24	15 hours
Iodine-131	8.07 days	Uranium-238	4.5 billion years

1. Use the data in the table above to complete the following chart.

Table of Remaining Radium

Number of years after formation	0	1600	3200	6400	12,800
Percent of radium-226 remaining	100%	50%			

2. If 1 g of sodium-24 has decayed from a sample that was originally 2 g, how old is the sample?

3. What fraction of chlorine-36 remains undecayed after 200,000 years?

Math Skills Used: Multiplication, Division, Fractions, Decimals, Percentages, Scientific Notation

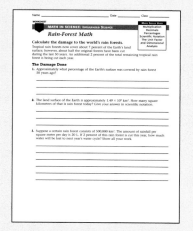

Rain-Forest Math

Calculate the damage to the world's rain forests.

Tropical rain forests now cover about 7 percent of the Earth's land surface; however, about half the original forests have been cut during the last 50 years. An additional 2 percent of the total remaining tropical rain forest is being cut each year.

The Damage Done

1. Approximately what percentage of the Earth's surface was covered by rain forest 50 years ago?

2. The land surface of the Earth is approximately 1.49 × 10⁸ km². How many square kilometers of that is rain forest today? Give your answer in scientific notation.

3. Suppose a certain rain forest consists of 500,000 km². The amount of rainfall per square meter per day is 20 L. If 2 percent of this rain forest is cut this year, how much water will be lost to next year's water cycle? Show all your work.

Math Skills Used: Multiplication, Decimals, Percentages, Scientific Notation, The Unit Factor and Dimensional Analysis

Math Skills for Science (continued)

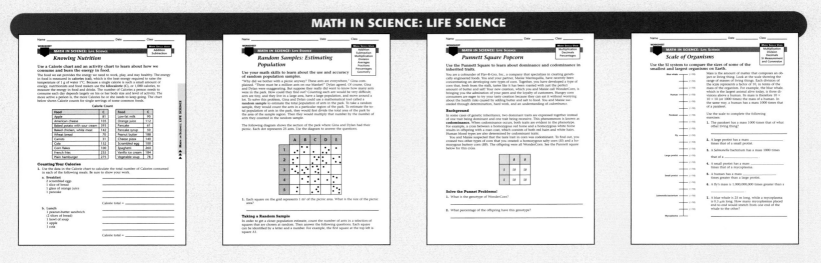

MATH IN SCIENCE: LIFE SCIENCE

Assessment Checklist & Rubrics

The following is just a sample of over 50 checklists and rubrics contained in this booklet.

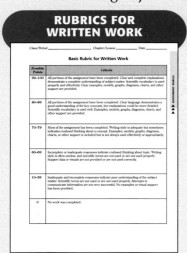

RUBRICS FOR WRITTEN WORK

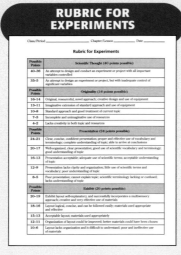

RUBRIC FOR EXPERIMENTS

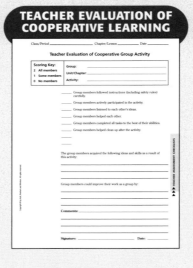

TEACHER EVALUATION OF COOPERATIVE LEARNING

TEACHER EVALUATION OF STUDENT PROGRESS

LIFE SCIENCE NATIONAL SCIENCE EDUCATION STANDARDS CORRELATIONS

The following lists show the chapter correlation of **Holt Science and Technology: Microorganisms, Fungi, and Plants** with the *National Science Education Standards* (grades 5-8)

UNIFYING CONCEPTS AND PROCESSES

Standard	Chapter Correlation	
Systems, order, and organization Code: UCP 1	Chapter 1 Chapter 2 Chapter 4 Chapter 5	1.1 2.1, 2.3 4.1, 4.2, 4.3, 4.4 5.2, 5.3
Evidence, models, and explanation Code: UCP 2	Chapter 1 Chapter 2 Chapter 3 Chapter 4 Chapter 5	1.1, 1.2, 1.3 2.2, 2.3 3.1 4.1, 4.2, 4.3, 4.4 5.1, 5.2, 5.3
Change, constancy, and measurement Code: UCP 3	Chapter 2 Chapter 4 Chapter 5	2.1, 2.2 4.2 5.1, 5.2, 5.3
Evolution and equilibrium Code: UCP 4	Chapter 2 Chapter 5	2.2 5.1, 5.2
Form and function Code: UCP 5	Chapter 1 Chapter 2 Chapter 3 Chapter 4 Chapter 5	1.3 2.1, 2.2, 2.3 3.1, 3.2 4.1, 4.2, 4.3, 4.4 5.1, 5.2

SCIENCE IN PERSONAL AND SOCIAL PERSPECTIVES

Standard	Chapter Correlation	
Personal health Code: SPSP 1	Chapter 2 Chapter 3	2.3 3.1
Populations, resources, and environments Code: SPSP 2	Chapter 5	5.3
Natural hazards Code: SPSP 3	Chapter 5	5.2, 5.3
Risks and benefits Code: SPSP 4	Chapter 2 Chapter 4 Chapter 5	2.3 4.3 5.1
Science and technology in society Code: SPSP 5	Chapter 2 Chapter 3 Chapter 4 Chapter 5	2.2, 2.3 3.2 4.4 5.3

SCIENCE AS INQUIRY

Standard	Chapter Correlation	
Abilities necessary to do scientific inquiry Code: SAI 1	Chapter 1 Chapter 2 Chapter 3 Chapter 4 Chapter 5	1.3 2.1, 2.2, 2.3 3.2 4.1, 4.2, 4.3, 4.4 5.1, 5.2, 5.3
Understandings about scientific inquiry Code: SAI 2	Chapter 1 Chapter 2 Chapter 3 Chapter 4 Chapter 5	1.2 2.1, 2.3 3.2 4.1 5.1, 5.2, 5.3

SCIENCE & TECHNOLOGY

Standard	Chapter Correlation	
Abilities of technological design Code: ST 1	Chapter 2	2.2
Understandings about science and technology Code: ST 2	Chapter 4	4.1

HISTORY AND NATURE OF SCIENCE

Standard	Chapter Correlation	
Science as a human endeavor Code: HNS 1	Chapter 1 Chapter 2 Chapter 3 Chapter 4 Chapter 5	1.2, 1.3 2.2 3.2 4.3, 4.4 5.1, 5.2
Nature of science Code: HNS 2	Chapter 1 Chapter 4 Chapter 5	1.2 4.1 5.2, 5.3
History of science Code: HNS 3	Chapter 2	2.2

LIFE SCIENCE

STRUCTURE AND FUNCTION IN LIVING SYSTEMS

Standard	Chapter Correlation	
Living systems at all levels of organization demonstrate the complementary nature of structure and function. Important levels of organization for structure and function include cells, organs, tissues, organ systems, whole organisms, and ecosystems. Code: LS 1a	**Chapter 1** **Chapter 4** **Chapter 5**	1.3 4.2, 4.3, 4.4 5.1, 5.2
All organisms are composed of cells—the fundamental unit of life. Most organisms are single cells; other organisms, including humans, are multicellular. Code: LS 1b	**Chapter 1** **Chapter 2** **Chapter 3**	1.2 2.1, 2.3 3.1, 3.2
Cells carry on the many functions needed to sustain life. They grow and divide, thereby producing more cells. This requires that they take in nutrients, which they use to provide energy for the work that cells do and to make the materials that a cell or an organism needs. Code: LS 1c	**Chapter 1** **Chapter 2** **Chapter 3** **Chapter 4** **Chapter 5**	1.1, 1.2, 1.3 2.1, 2.2 3.1, 3.2 4.2 5.2
Specialized cells perform specialized functions in multicellular organisms. Groups of specialized cells co-operate to form a tissue, such as a muscle. Different tissues are in turn grouped together and form larger functional units, called organs. Each type of cell, tissue, and organ has a distinct structure and set of functions that serve the organism as a whole. Code: LS 1d	**Chapter 1** **Chapter 3** **Chapter 4**	1.1 3.2 4.3, 4.4
Disease is a breakdown in structures or functions of an organism. Some diseases are the result of intrinsic failures of the system. Others are the result of damage by infection by other organisms. Code: LS 1f	**Chapter 2** **Chapter 3**	2.2, 2.3 3.1, 3.2

REPRODUCTION AND HEREDITY

Standard	Chapter Correlation	
Reproduction is a characteristic of all living systems; because no individual organism lives forever, reproduction is essential to the continuation of every species. Some organisms reproduce asexually. Others reproduce sexually. Code: LS 2a	**Chapter 1** **Chapter 2** **Chapter 3** **Chapter 5**	1.1 2.1 3.1, 3.2 5.1
In many species, including humans, females produce eggs and males produce sperm. Plants also reproduce sexually—the egg and sperm are produced in the flowers of flowering plants. An egg and sperm unite to begin development of a new individual. The individual receives genetic information from its mother (via the egg) and its father (via the sperm). Sexually produced offspring never are identical to either of their parents. Code: LS 2b	**Chapter 1** **Chapter 4** **Chapter 5**	1.1 4.1, 4.2, 4.3, 4.4 5.1, 5.3
Every organism requires a set of instructions for specifying its traits. Heredity is the passage of these instructions from one generation to another. Code: LS 2c	**Chapter 1** **Chapter 4** **Chapter 5**	1.1, 1.3 4.3 5.3
Hereditary information is contained in the genes, located in the chromosomes of each cell. Each gene carries a single unit of information. An inherited trait of an individual can be determined by one or by many genes, and a single gene can influence more than one trait. A human cell contains many thousands of different genes. Code: LS 2d	**Chapter 4** **Chapter 5**	4.3 5.1

Standard	Chapter Correlation	
All organisms must be able to obtain and use resources, grow, reproduce, and maintain stable internal conditions while living in a constantly changing external environment. Code: LS 3a	**Chapter 1** **Chapter 2** **Chapter 3** **Chapter 5**	1.1, 1.2 2.1 3.1 5.3
Regulation of an organism's internal environment involves sensing the internal environment and changing physiological activities to keep conditions within the range required to survive. Code: LS 3b	**Chapter 1** **Chapter 2**	1.1 2.1
Behavior is one kind of response an organism can make to an internal or environmental stimulus. A behavioral response requires coordination and communication at many levels, including cells, organ systems, and whole organisms. Behavioral response is a set of actions determined in part by heredity and in part from experience. Code: LS 3c	**Chapter 1** **Chapter 5**	1.1, 1.3 5.2, 5.3
An organism's behavior evolves through adaptation to its environment. How a species moves, obtains food, reproduces, and responds to danger are based in the species' evolutionary history. Code: LS 3d	**Chapter 1** **Chapter 4** **Chapter 5**	1.2 4.4 5.3

Standard	Chapter Correlation	
Populations of organisms can be categorized by the functions they serve in an ecosystem. Plants and some microorganisms are producers—they make their own food. All animals, including humans, are consumers, which obtain food by eating other organisms. Decomposers, primarily bacteria and fungi, are consumers that use waste materials and dead organisms for food. Food webs identify the relationships among producers, consumers, and decomposers in an ecosystem. Code: LS 4b	**Chapter 1** **Chapter 2** **Chapter 3** **Chapter 4**	1.2 2.1 3.1, 3.2 4.3
For ecosystems, the major source of energy is sunlight. Energy entering ecosystems as sunlight is transferred by producers into chemical energy through photosynthesis. That energy then passes from organism to organism in food webs. Code: LS 4c	**Chapter 1** **Chapter 4** **Chapter 5**	1.2 4.1, 4.3, 4.4 5.2
The number of organisms an ecosystem can support depends on the resources available and abiotic factors, such as quantity of light and water, range of temperatures, and soil composition. Given adequate biotic and abiotic resources and no disease or predators, populations (including humans) increase at rapid rates. Lack of resources and other factors, such as predation and climate, limit the growth of populations in specific niches in the ecosystem. Code: LS 4d	**Chapter 1** **Chapter 4**	1.2 4.3

Standard	Chapter Correlation	
Millions of species of animals, plants, and microorganisms are alive today. Although different species might look dissimilar, the unity among organisms becomes apparent from an analysis of internal structures, the similarity of their chemical processes, and the evidence of common ancestry. Code: LS 5a	**Chapter 1** **Chapter 3** **Chapter 4**	1.1 3.1 4.1
Biological evolution accounts for the diversity of species developed through gradual processes over many generations. Species acquire many of their unique characteristics through biological adaptation, which involves the selection of naturally occurring variations in populations. Biological adaptations include changes in structures, behaviors, or physiology that enhance survival and reproductive success in a particular environment. Code: LS 5b	**Chapter 4** **Chapter 5**	4.3, 4.4 5.1, 5.3
Extinction of a species occurs when the environment changes and the adaptive characteristics of a species are insufficient to allow its survival. Fossils indicate that many organisms that lived long ago are extinct. Extinction of species is common; most of the species that have lived on Earth no longer exist. Code: LS 5c	**Chapter 4**	4.2

Master Materials List

For added convenience, Science Kit® provides materials-ordering software on CD-ROM designed specifically for *Holt Science and Technology*. Using this software, you can order complete kits or individual items, quickly and efficiently.

CONSUMABLE MATERIALS	AMOUNT	PAGE
Agar, nutrient	30mL	23
Aluminum foil	1 sheet	73
Bag, plastic, sealable	1	59
Baking soda	25g	118
Bottle, soda, 2 L	1	73
Bread, slice	1	59
Cardboard, 2 x 2 ft	1	114
Chalk	1 stick	14
Cup, plastic	3	134
Cup, plastic	1	105
Elodea sprig, 20 cm long	3	118
Flour	1/2 cup	134
Food sample, various types	5	11
Fruit-juice agar plate	1	64
Gloves, protective	1 pair	11, 14, 96, 118, 130, 136
Glue, white	1 bottle	36, 137
Hay infusion	1 drop	45
Ice		132
Iodine solution	20 mL	11
Isopod	4	14
Leaf, fresh, various kinds	5	96
Marker, black, nonpermanent	1	130, 138
Marker, black, permanent	1	105
Marker, various colors	1 pack	36
Moss, dry sphagnum	1 sample	79
Mushroom	1	61, 64
Newspaper	1	133
Paper, construction, various colors	5 sheets	36
Paper, graph	1 sheet	132, 138

CONSUMABLE MATERIALS	AMOUNT	PAGE
Paper, white	2 sheets	64
Paper towel	1 roll	105
Pencil, wax	1	108
Pipe cleaner	3	36
Plant, potted	3	114
Plant, stem cutting	1	138
Plastic wrap, clear, 3 x 10 cm	1	133
Plastic wrap, clear, approx. 1 x 2 ft	1 sheet	36
Pond water with living organisms	300 mL	45
Poster board, 3 x 10 cm	1	133
Potato, small slice	1	14
Proto Slo™	5 drops	45
Rubbing alcohol	75 mL	130
Seed, bean	1	137
Seed, bean	12	108
Seed, corn	6	105
Seed, kidney bean	4	73
Soil	8 oz	14, 23
Soil, potting	4 cups	73
String (or yarn)	3 m	36
Sugar, granulated	1/4 cup	134
Tape, duct	1 m	114
Tape, masking	25 cm	64
Tape, transparent	4 cm	133
Tape, transparent	1 roll	23, 36, 64
Yeast, active, dry-baking	1 packet	134
Yeast, active or inactive, dry-baking	1 packet	134
Yeast, inactive, dry-baking	1 packet	134

Nonconsumable Equipment	Amount	Page
Beaker, 250 mL	1	134
Beaker, 400 mL	1	132
Beaker, 600 mL	1	79, 118
Container, plastic with lid	1	14
Coverslip, plastic	1	45
Eyedropper, plastic	1	45, 133
Flashlight	1	3
Funnel, glass	1	118
Gloves, heat-resistant	1 pair	132, 134
Graduated cylinder, 50 mL	1	134
Graduated cylinder, 100 mL	7	130
Hole punch	1	133
Hot plate	1	132, 134
Incubator	1	64
Knife, plastic	1	61
Magnifying lens	1	61, 64, 134
Microscope, compound	1	26, 45, 64
Microscope slide, plastic	1	45

Nonconsumable Equipment	Amount	Page
Petri dish, plastic	1	64
Petri dish, plastic	2	108
Petri dish, plastic	3	23
Plant guidebook	1	96
Ruler, metric	1	14, 118, 138
Scissors	1	36
Scoopula	1	134
Slide, prepared, bacteria	3	26
Stirring stick, wooden	3	134
Stopwatch	1	14, 132, 138
Test tube	1	118
Test tube	2	138
Test tube	3	134
Test-tube rack	1	134, 138
Thermometer, Celsius	1	130, 132, 134
Thermometer clip	1	132, 134
Tweezers	1	64

Answers to Concept Mapping Questions

The following pages contain sample answers to all of the concept mapping questions that appear in the Chapter Reviews. Because there is more than one way to do a concept map, your students' answers may vary.

CHAPTER 1 It's Alive!! Or, Is It?

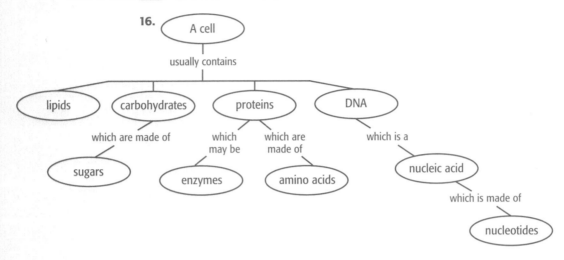

16.

CHAPTER 2 Bacteria and Viruses

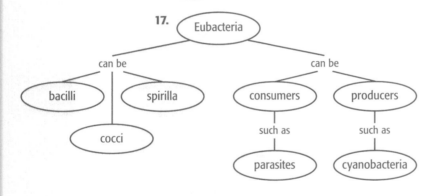

17.

CHAPTER 3 Protists and Fungi

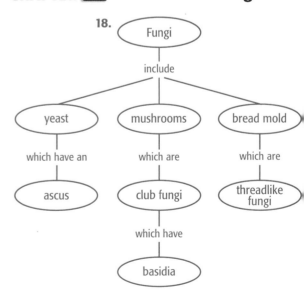

18.

17.

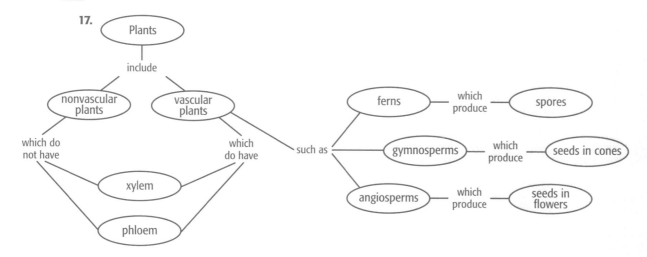

CHAPTER 5 Plant Processes

15.

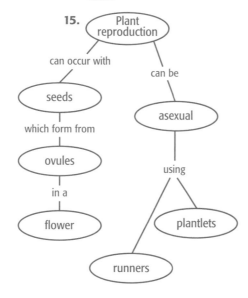

To the Student

This book was created to make your science experience interesting, exciting, and fun!

Go for It!

Science is a process of discovery, a trek into the unknown. The skills you develop using *Holt Science & Technology*—such as observing, experimenting, and explaining observations and ideas—are the skills you will need for the future. There is a universe of exploration and discovery awaiting those who accept the challenges of science.

Science & Technology

You see the interaction between science and technology every day. Science makes technology possible. On the other hand, some of the products of technology, such as computers, are used to make further scientific discoveries. In fact, much of the scientific work that is done today has become so technically complicated and expensive that no one person can do it entirely alone. But make no mistake, the creative ideas for even the most highly technical and expensive scientific work still come from individuals.

Activities and Labs

The activities and labs in this book will allow you to make some basic but important scientific discoveries on your own. You can even do some exploring on your own at home! Here's your chance to use your imagination and curiosity as you investigate your world.

Keep a ScienceLog

In this book, you will be asked to keep a type of journal called a ScienceLog to record your thoughts, observations, experiments, and conclusions. As you develop your ScienceLog, you will see your own ideas taking shape over time. You'll have a written record of how your ideas have changed as you learn about and explore interesting topics in science.

Know "What You'll Do"

The "What You'll Do" list at the beginning of each section is your built-in guide to what you need to learn in each chapter. When you can answer the questions in the Section Review and Chapter Review, you know you are ready for a test.

Check Out the Internet

You will see this $^{sc}_{LINKS}$ logo throughout the book. You'll be using *sci*LINKS as your gateway to the Internet. Once you log on to *sci*LINKS using your computer's Internet link, type in the *sci*LINKS address. When asked for the keyword code, type in the keyword for that topic. A wealth of resources is now at your disposal to help you learn more about that topic.

In addition to *sci*LINKS you can log on to some other great resources to go with your text. The addresses shown below will take you to the home page of each site.

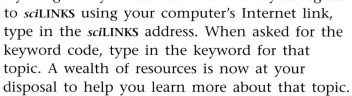

internet**connect**

This textbook contains the following on-line resources to help you make the most of your science experience.

go.hrw.com	SCiLINKS NSTA	Smithsonian Institution® Internet Connections	CNNfyi.com
Visit **go.hrw.com** for extra help and study aids matched to your textbook. Just type in the keyword HG2 HOME.	Visit **www.scilinks.org** to find resources specific to topics in your textbook. Keywords appear throughout your book to take you further.	Visit **www.si.edu/hrw** for specifically chosen on-line materials from one of our nation's premier science museums.	Visit **www.cnnfyi.com** for late-breaking news and current events stories selected just for you.

Chapter Organizer

CHAPTER ORGANIZATION	TIME MINUTES	OBJECTIVES	LABS, INVESTIGATIONS, AND DEMONSTRATIONS
Chapter Opener pp. 2–3	45	National Standards: UCP 5, SAI 1, LS 3c, 4c	**Start-Up Activity,** Lights On! p. 3
Section 1 **Characteristics of Living Things**	90	▶ List the characteristics of living things. ▶ Distinguish between asexual reproduction and sexual reproduction. ▶ Define and describe homeostasis. UCP 1, 2, LS 1b–1d, 2a–2c, 3a–3c, 5a	**Discovery Lab,** Roly-Poly Races, p. 14 **Datasheets for LabBook,** Roly-Poly Races
Section 2 **The Simple Bare Necessities of Life**	90	▶ Explain why organisms need food, water, air, and living space. ▶ Discuss how living things obtain what they need to live. UCP 2, SAI 2, HNS 1, 2, LS 1c, 3a, 3d, 4b–4d	**Demonstration,** Fire and Life, p. 9 in ATE
Section 3 **The Chemistry of Life**	90	▶ Compare and contrast the chemical building blocks of cells. ▶ Explain the importance of ATP. UCP 2, 5, SAI 1, LS 1a, 1c, 2c; Labs UCP 2, SAI 1, HNS 1, LS 3c	**Demonstration,** Protein Model, p. 10 in ATE **QuickLab,** Starch Search, p. 11 **Discovery Lab,** The Best-Bread Bakery Dilemma, p. 134 **Datasheets for LabBook,** The Best-Bread Bakery Dilemma **Labs You Can Eat,** Say Cheese! **Long-Term Projects & Research Ideas,** I Think, Therefore I Live

*See page **T23** for a complete correlation of this book with the*

NATIONAL SCIENCE EDUCATION STANDARDS.

TECHNOLOGY RESOURCES

 Guided Reading Audio CD
English or Spanish, Chapter 1

 One-Stop Planner CD-ROM with Test Generator

 CNN **Science, Technology & Society,** Tapping into Yellowstone's Hot Springs, Segment 1

Chapter 1 • It's Alive!! Or, Is It?

CLASSROOM WORKSHEETS, TRANSPARENCIES, AND RESOURCES	SCIENCE INTEGRATION AND CONNECTIONS	REVIEW AND ASSESSMENT
Directed Reading Worksheet **Science Puzzlers, Twisters & Teasers**		
Directed Reading Worksheet, Section 1 **Math Skills for Science Worksheet,** A Shortcut for Multiplying Large Numbers **Math Skills for Science Worksheet,** Multiplying and Dividing in Scientific Notation **Critical Thinking Worksheet,** Intergalactic Planetary Mission **Math Skills for Science Worksheet,** Decimals and Fractions **Math Skills for Science Worksheet,** Percentages, Fractions, and Decimals	**Oceanography Connection,** p. 5 **Math and More,** p. 5 in ATE **Math and More,** p. 6 in ATE **Apply,** p. 7	**Self-Check,** p. 5 **Section Review,** p. 7 **Quiz,** p. 7 in ATE **Alternative Assessment,** p. 7 in ATE
Directed Reading Worksheet, Section 2 **Transparency 178,** Climate Zones of the Earth **Reinforcement Worksheet,** Amazing Discovery	**Connect to Earth Science,** p. 8 in ATE	**Section Review,** p. 9 **Quiz,** p. 9 in ATE **Alternative Assessment,** p. 9 in ATE
Directed Reading Worksheet, Section 3 **Transparency 5,** Phospholipid Molecule and Cell Membrane **Transparency 6,** Energy for Cells **Reinforcement Worksheet,** Building Blocks	**Multicultural Connection,** p. 10 in ATE **MathBreak,** How Much Oxygen? p. 11 **Scientific Debate:** Life on Mars? p. 20 **Holt Anthology of Science Fiction,** *They're Made Out of Meat*	**Homework,** p. 12 in ATE **Section Review,** p. 13 **Quiz,** p. 13 in ATE **Alternative Assessment,** p. 13 in ATE

 internet**connect**

 go. hrw .com **Holt, Rinehart and Winston On-line Resources**

go.hrw.com

For worksheets and other teaching aids related to this chapter, visit the HRW Web site and type in the keyword: **HSTALV**

 SC*i*LINKS NSTA **National Science Teachers Association**

www.scilinks.org

Encourage students to use the *sci*LINKS numbers listed in the internet connect boxes to access information and resources on the **NSTA** Web site.

END-OF-CHAPTER REVIEW AND ASSESSMENT

Chapter Review in Study Guide
Vocabulary and Notes in Study Guide
Chapter Tests with Performance-Based Assessment, Chapter 1 Test
Chapter Tests with Performance-Based Assessment, Performance-Based Assessment 1
Concept Mapping Transparency 2

Chapter Resources & Worksheets

Visual Resources

TEACHING TRANSPARENCIES

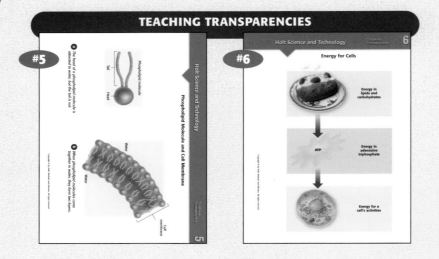

#5

Holt Science and Technology

Phospholipid Molecule and Cell Membrane

#6

Holt Science and Technology

Energy for Cells

- Energy in lipids and carbohydrates
- ATP — Energy in adenosine triphosphate
- Energy for a cell's activities

TEACHING TRANSPARENCIES

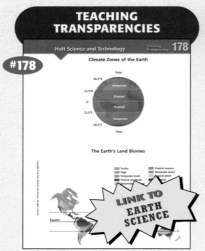

#178

Holt Science and Technology

Teaching Transparency 178

Climate Zones of the Earth

Polar
66.5°N
Temperate
23.5°N
Tropical
0°
Tropical
23.5°S
Temperate
66.5°S
Polar

The Earth's Land Biomes

LINK TO EARTH SCIENCE

Equator

CONCEPT MAPPING TRANSPARENCY

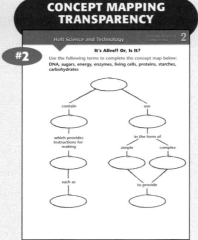

#2

Holt Science and Technology

Concept Mapping Transparency 2

It's Alive!! Or, Is It?

Use the following terms to complete the concept map below:
DNA, sugars, energy, enzymes, living cells, proteins, starches, carbohydrates

contain — use

which provides instructions for making

in the form of

simple — complex

such as

to provide

Meeting Individual Needs

DIRECTED READING

#1

Date _____ Class _____

DIRECTED READING WORKSHEET

It's Alive!! Or, Is It?

Chapter Introduction

As you begin this chapter, answer the following.
1. Read the title of the chapter. List three things that you already know about this subject.

2. Write two questions about this subject that you would like answered by the time you finish this chapter.

3. How does the title of the Start-Up Activity relate to the subject of the chapter?

Section 1: Characteristics of Living Things (p. 4)

4. What might you have in common with a slime mold?

5. All organisms, including fish, trees, and mushrooms, share _____ characteristics.

REINFORCEMENT & VOCABULARY REVIEW

#1

Date _____ Class _____

REINFORCEMENT WORKSHEET

Amazing Discovery

Complete this worksheet after you finish reading Chapter 2, Section 2.
Imagine that you are a biologist on a mission to Mars. You have just discovered what you think is a simple one-celled Martian organism. For now, you are calling it Alpha. Before you can claim that you have discovered life on Mars, however, you need to prove that Alpha is alive.

1. What are the six characteristics you will look for to see if Alpha is alive?
   a.
   b.
   c.
   d.
   e.

2. Outline a test or experiment to verify one of the characteristics you listed above.

3. If you can prove that Alpha is alive, you will take it back to Earth for further study. What will you need to provide Alpha with to keep it alive?

#1

Date _____ Class _____

VOCABULARY REVIEW WORKSHEET

It's Alive!

Complete this puzzle after you finish Chapter 2.
In the space provided, write the term described by the clue. Then find these words in the puzzle. Terms can be hidden in the puzzle vertically, horizontally, diagonally or backwards.

1. _____ change in an organism's environment that affects the activity of an organism
2. _____ group of compounds made of sugars
3. _____ maintenance of a stable internal environment
4. _____ complex carbohydrate made by plants
5. _____ transmission of characteristics from one generation to the next
6. _____ chemical activities of an organism necessary for life
7. _____ made up of subunits called nucleotides
8. _____ eats other organisms for food
9. _____ organism that breaks down the nutrients of dead organisms or wastes for food
10. _____ two layers of these form much of the cell membrane
11. _____ proteins that speed up certain chemical reactions
12. _____ molecule that provides instructions for making proteins
13. _____ organism that can produce its own food
14. _____ membrane-covered structure that contains all materials necessary for life
15. _____ reproduction in which a single parent produces offspring that are identical to the parent
16. _____ chemical compound that cannot mix with water and that is used to store energy
17. _____ large molecule made up of amino acids
18. _____ energy in food is transferred to this molecule
19. _____ reproduction in which two parents are necessary to produce offspring that share characteristics of both parents

SCIENCE PUZZLERS, TWISTERS & TEASERS

#1

Date _____ Class _____

SCIENCE PUZZLERS, TWISTERS & TEASERS

It's Alive!! Or, Is It?

Blocks of Life

1. Sean borrowed his baby sister's blocks to help him prepare for a vocabulary test in biology. He arranged them so that each row of blocks below spells the name of a substance that is a building block of cells. But while he wasn't looking, his sister rotated some of the blocks so that the wrong side is facing toward the front. Choose one letter from each of the blocks in a row to spell the names of some important compounds that are found in cells. Record the names of the compounds on the blanks provided.

   a. _____
   b. _____
   c. _____

Life: Finding the Right Combination

2. Six words related to the six characteristics of living organisms are hidden in the circle below. Rotate each ring of the circle so that each pie-shaped section of the circle spells one of the words as you read from the outside ring and move toward the inside ring. Write the words on the spaces provided.

Chapter 1 • It's Alive!! Or, Is It?

Review & Assessment

STUDY GUIDE

#1 VOCABULARY & NOTES WORKSHEET
It's Alive!! Or, Is It?

By studying the Vocabulary and Notes listed for each section below, you can gain a better understanding of this chapter.

SECTION 1
Vocabulary
In your own words, write a definition for each of the following terms in the space provided.

1. cell
2. stimulus
3. homeostasis
4. asexual reproduction
5. sexual reproduction
6. DNA
7. heredity
8. metabolism

#1 CHAPTER REVIEW WORKSHEET
It's Alive!! Or, Is It?

USING VOCABULARY
To complete the following sentences, choose the correct terms from each set of terms listed below, and write the term in the space provided.
1. The process of maintaining a stable internal environment is known as ___. (metabolism or homeostasis)
2. The resemblance of offspring to their parents is a result of ___. (heredity or stimuli)
3. A ___ obtains food by eating other organisms. (producer or consumer)
4. Starch is a ___ and is made up of ___. (carbohydrate/sugars or nucleic acid/nucleotides)
5. Fats and oils are ___ that store energy for an organism. (proteins or lipids)

UNDERSTANDING CONCEPTS
Multiple Choice
6. Cells are
   a. the structures that contain all the materials necessary for life.
   b. found in all organisms.
   c. sometimes specialized for particular functions.
   d. All of the above
7. Which of the following is a true statement about all living things?
   a. They cannot sense changes in their external environment.
   b. They have one or more cells.
   c. They do not need to use energy.
   d. They reproduce asexually.
8. Organisms must have food because
   a. food is a source of energy.
   b. food supplies cells with oxygen.
   c. organisms never make their own food.
   d. All of the above
9. A change in an organism's environment that affects the organism's activities is a
   a. response.
   b. stimulus.
   c. metabolism.
   d. producer.

CHAPTER TESTS WITH PERFORMANCE-BASED ASSESSMENT

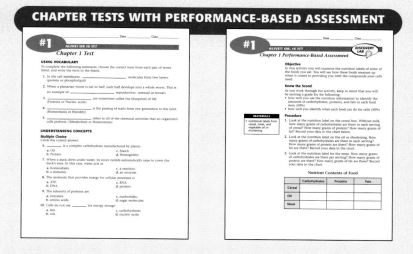

#1 ALIVE!! OR IS IT?
Chapter 1 Test

USING VOCABULARY
To complete the following sentences, choose the correct term from each pair of terms listed, and write the term in the blank.
1. In the cell membrane, ___ molecules form two layers. (protein or phospholipid)
2. When a planarian worm is cut in half, each half develops into a whole worm. This is an example of ___ reproduction. (asexual or sexual)
3. ___ are sometimes called the blueprints of life. (Proteins or Nucleic acids)
4. ___ is the passing of traits from one generation to the next. (Homeostasis or Heredity)
5. ___ refers to all of the chemical activities that an organism's cells perform. (Metabolism or Homeostasis)

UNDERSTANDING CONCEPTS
Multiple Choice
Circle the correct answer.
6. ___ is a complex carbohydrate manufactured by plants.
   a. Oil   c. Starch
   b. Protein   d. Hemoglobin
7. When a duck dives under water, its inner eyelids automatically raise to cover the duck's eyes. In this case, water acts as
   a. homeostasis.   c. a reaction.
   b. a stimulus.   d. an enzyme.
8. The molecule that provides energy for cellular processes is
   a. ATP.   c. RNA.
   b. DNA.   d. protein.
9. The subunits of proteins are
   a. enzymes.   c. nucleotides.
   b. amino acids. d. sugar molecules.
10. Cells do not use ___ for energy storage.
   a. fats   c. carbohydrates
   b. oils   d. nucleic acids

#1 ALIVE!! OR IS IT?   DISCOVERY LAB
Chapter 1 Performance-Based Assessment

Objective
In this activity you will examine the nutrition labels of some of the foods you eat. You will see how these foods measure up when it comes to providing you with the compounds your cells need.

Know the Score!
As you work through the activity, keep in mind that you will be earning a grade for the following:
• how well you use the nutrition information to identify the amounts of carbohydrates, proteins, and fats in each food item (50%)
• how well you identify what each food can do for cells (50%)

MATERIALS
• nutritional labels from cereal, meat, and vegetable oil or shortening

Procedure
1. Look at the nutrition label on the cereal box. Without milk, how many grams of carbohydrates are there in each serving of cereal? How many grams of protein? How many grams of fat? Record your data in the chart below.
2. Look at the nutrition label on the oil or shortening. How many grams of carbohydrates are there in each serving? How many grams of protein are there? How many grams of fat are there? Record your data in the chart.
3. Look at the nutrition label for the meat. How many grams of carbohydrates are there per serving? How many grams of protein are there? How many grams of fat are there? Record your data in the chart.

Nutrient Contents of Food

	Carbohydrates	Proteins	Fats
Cereal			
Oil			
Meat			

Lab Worksheets

LABS YOU CAN EAT

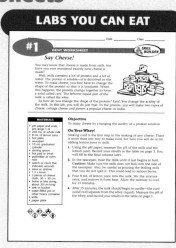

#1 STUDENT WORKSHEET   SKILL BUILDER
Say Cheese!

You may know that cheese is made from milk, but have you ever wondered exactly how cheese is made?

Well, milk contains a lot of protein and a lot of water. The protein is soluble—it is dissolved in the water. To make cheese, you first have to change the shape of the protein so that it is insoluble. When this happens, the protein clumps together to form a solid called *curd*. The leftover liquid part of the milk is called the *whey*.

So how do you change the shape of the protein? Easy! You change the acidity of the milk. In this lab, you will do just that. In the process, you will make two types of cheese: cottage cheese and *paneer*, a popular cheese in India.

MATERIALS
• pH paper and scale, pH range 1–8
• 250 mL of whole milk
• 8 mL of lemon juice
• hot plate
• saucepan
• 10 mL graduated cylinder
• stirring spoon
• hot pad or trivet
• potholder or oven mitt
• watch or clock that indicates seconds
• sieve or strainer
• 1.5-L bowl
• 3 pieces of cheese cloth, 30 × 30 cm
• twine or kite string about 20 cm long
• sink or bucket
• water-filled pot or other heavy container
• knife
• paper plate

Objective
To make cheese by changing the acidity of a protein solution.

On Your Whey!
Making curd is the first step in the making of any cheese. There is more than one way to make curd, but here you will do so by adding lemon juice to milk.
1. Using the pH paper, measure the pH of the milk and the lemon juice. Record your results in the table on page 3. You will fill in the final column later.
2. In the saucepan, heat the milk until it just begins to boil. **Caution:** Make sure the milk does not boil over the side of the saucepan. Also, be careful in pouring the boiling milk that you do not spill it. This could lead to serious burns.
3. Pour 8 mL of lemon juice into the milk. Stir the mixture once, and remove it from heat. Allow the mixture to cool for 15 minutes.
4. After 15 minutes, the milk should begin to curdle—the curd (solid) will separate from the whey (liquid). Measure the pH of the whey, and record your results in the table on page 3.

LONG-TERM PROJECTS & RESEARCH IDEAS

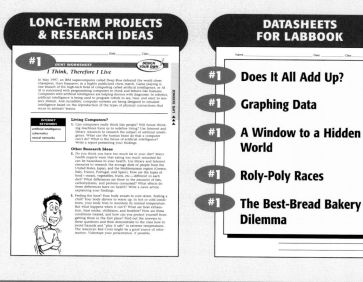

#1 STUDENT WORKSHEET   DESIGN YOUR OWN
I Think, Therefore I Live

In May 1997, an IBM supercomputer called Deep Blue defeated the world chess champion, Gary Kasparov, in a highly publicized chess match. Game playing is one branch of the high-tech field of computing called artificial intelligence, or AI. AI is concerned with programming computers to think and behave like humans. Computers with artificial intelligence are helping doctors with diagnoses. In robotics, artificial intelligence is being used to program robots to see, hear, and react to sensory stimuli. And incredibly, computer systems are being designed to simulate intelligence based on the reproduction of the types of physical connections that occur in animals' brains.

INTERNET KEYWORDS
artificial intelligence
cybernetics
neural networks

Living Computers?
1. Can computers really think like people? Will future thinking machines force us to redefine living? Use Internet and library resources to research the subject of artificial intelligence. What can the human brain do that a computer can't do? What is the future of artificial intelligence? Write a report presenting your findings.

Other Research Ideas
2. Do you think you have too much fat in your diet? Many health experts warn that eating too much saturated fat can be hazardous to your health. Use library and Internet resources to research the average diets of people from the United States, Japan, and the Mediterranean region (Greece, Italy, France, Portugal, and Spain). How are the types of food—meats, vegetables, fruits, etc.—different in each diet? What differences are there in the amounts of fats, carbohydrates, and proteins consumed? What effects do these differences have on health? Write a news article explaining your findings.
3. Feeling the heat? Your body sweats to cool down. Feeling a chill? Your body shivers to warm up. In hot or cold conditions, your body tries to maintain its normal temperature. But what happens when it can't? What are heat exhaustion, heat stroke, chilblains, and frostbite? How are these conditions treated, and how can you protect yourself from getting them in the first place? Find out the answers to these questions and then demonstrate to the class how to avoid hazards and "play it safe" in extreme temperatures. The American Red Cross might be a good source of information. Videotape your presentation, if possible.

DATASHEETS FOR LABBOOK

#1 Does It All Add Up?

#1 Graphing Data

#1 A Window to a Hidden World

#1 Roly-Poly Races

#1 The Best-Bread Bakery Dilemma

Applications & Extensions

CRITICAL THINKING & PROBLEM SOLVING

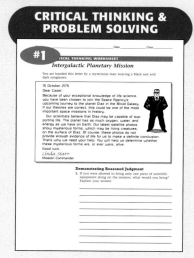

#1 CRITICAL THINKING WORKSHEET
Intergalactic Planetary Mission

You may have received this letter by a mysterious man wearing a black suit and dark sunglasses:

15 October 2075
Dear Cadet:
Because of your exceptional knowledge of life science, you have been chosen to join the Space Agency's upcoming journey to the planet Diaz in the Blixil Galaxy. If our theories are correct, this could be one of the most important space missions in history.
Our scientists believe that Diaz may be capable of supporting life. The planet has as much oxygen, water, and energy as we have on Earth. Our latest satellite photos show mysterious forms, which may be living creatures, on the surface of Diaz. Of course, these photos do not provide enough evidence of life for us to make a definite conclusion. That's why we need your help. You will help us determine whether these mysterious forms are, or ever were, alive.
Good luck.
Linda Starr
Mission Commander

Demonstrating Reasoned Judgment
1. If you were allowed to bring only one piece of scientific equipment along on the mission, what would you bring? Explain your answer.

SCIENCE TECHNOLOGY

#1 Science in the News: Critical Thinking Worksheets
Segment 1
Tapping into Yellowstone's Hot Springs

1. Why do you think scientists believe that hot-water life-forms may have lived on the Earth before other types of life-forms?

2. How do you think scientists study microscopic life-forms that live deep within the Earth?

3. What impact could the disappearance of these bacteria have on humans?

4. Who do you think should fund the protection of ...

Chapter Background

Characteristics of Living Things

▶ Biogenesis

The theory of biogenesis states that living things come only from other living things. However, until the late 1600s, people generally believed in spontaneous genera-tion, the theory that lower forms of 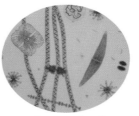 life, such as insects, come from nonliving things. The first evidence to disprove spontaneous generation came from controlled experiments conducted in 1667 by Italian Francesco Redi.

- Redi showed that maggots will appear on meat in an uncovered jar but not on meat in a closed container. Why? The maggots came from eggs laid by flies that had access to the uncovered meat.

▶ Robert Hooke

Robert Hooke was one of the greatest scientists of his time. In addition to his studies in biology with the compound microscope, he was also involved in physics, astronomy, chemistry, geology, and architecture.

- Hooke applied his discovery of the law of elasticity (that the stretching of a solid material is proportional to the force applied to it) to the design of balance springs for watches and clocks. His sketches of Mars were used 200 years later to determine that planet's rate of rotation. He studied the crystal structure of snowflakes. In 1672, he developed the wave theory of light to explain diffraction, which he had also discov-ered. Hooke was the first person to examine fossils with a microscope and to recognize, 200 years before Charles Darwin was born, that fossils are evidence of changes in organisms on Earth over millions of years.

IS THAT A FACT!

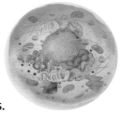

- ➧ Different cells in the human body have different life spans, ranging from a few days for intestinal cells to about 120 days for red blood cells and years for brain cells. Scientists have recently discovered that the brain does make new cells.

However, they only mature in the hippocampus, which is responsible for learning and memory.

▶ Anton van Leeuwenhoek

Anton van Leeuwenhoek was born in 1632 into a family of tradespeople in Delft, Holland. He was poor and received no advanced education. In 1648, he was apprenticed to a fabric merchant to learn this trade. On his own initiative, he began to grind lenses to experiment with magnification. Historians think he may have been inspired by Robert Hooke's book, *Micrographia* (Small Drawings).

The Simple Bare Necessities of Life

▶ A Place to Call Home

Every organism needs a place to live, and the places where some organisms thrive are surprising.

- Antarctica is a harsh and unforgiving environment. Most of the subantarctic islands are solid rock, and 98 percent of the continent is covered with ice. But it is home to more than 400 types of lichens and 85 mosses. Lichens, a symbiotic association of algae and fungi, can tolerate low temperatures and minimal moisture. Green algae thrive near penguin colonies. Moss grows on the few small patches of soil that exist.

- In Death Valley National Park, summertime tempera-tures routinely reach 50°C, and rainfall averages 3.8 cm per year. Yet more than 900 types of plants live there. More than 400 animal species also live in this region, including bats, kangaroo rats, gophers, bighorn sheep, lizards, tortoises, snakes, spiders, scorpions, beetles, turkey vultures, and roadrunners.

- Three to four kilometers deep in the ocean, where the pressure is 275 times that at sea level and it is cold and dark, lurk huge yellow jellyfish, giant clams, blind fish, and red worms that are 2 m long. These animals live near deep-sea vents. Water that is trapped beneath the ocean floor and is heated by volcanic activity to as much as 300°C escapes through these vents. The cold ocean cools the water around the vents to about 13°C. Because there is no sunlight at this depth, the animals use the chemicals in the water for energy through a process called chemosynthesis. However, the amount of energy available through chemosynthesis is far less than that obtained through photosynthesis on Earth's surface.

IS THAT A FACT!

- The sand grouse of Chad, in northern Africa, builds its nest many miles from a water source. When the chicks hatch, the parents fly to Lake Chad, where they soak their breast feathers before flying back to their chicks. The chicks then drink the water from their parents' feathers; this both feeds and cools them.

- Some animals that hunt at night, such as cats and some dogs, have a special eye structure that helps them see in the dark. The *tapetum lucidum,* or "bright carpet," a mirrorlike layer of cells, enhances the eye's ability to see in low light. It makes a cat's eyes appear to glow in the dark.

SECTION 3

The Chemistry of Life

▶ "Chemical" Menu
Our bodies need certain kinds of chemicals to live.

- Carbohydrates are the body's primary source of energy. Simple carbohydrates, or sugars, are found in fruits, some vegetables, and milk. Complex carbohydrates, which include starches, are obtained from pasta, seeds, nuts, and vegetables such as peas, beans, and potatoes.

- Lipids include saturated and polyunsaturated fats. Saturated fats are present in greater amounts in animal products. Vegetable-based oils have more polyunsaturated fats.

- Nucleic acids, which may contain thousands of components called nucleotides, contain essential information for the construction of proteins.

- Proteins are made of 20 different amino acids. Our cells arrange these amino acids in different combinations to make all the proteins in our body.

IS THAT A FACT!

- If stretched out end to end, the DNA in an average human body would measure 20 billion kilometers.

- For about 100 years, beginning in the late 1700s, sperm whales were a major source of oil for lubricants and fuel for lamps. These huge animals grow to 18 m long. Sperm oil came from the whale's blubber and unusually large head. Whaling nearly made sperm whales extinct.

▶ Metabolism
Biochemical reactions that take place within a cell are collectively known as metabolism. Enzymes, which are themselves proteins, catalyze or accelerate most of the chemical reactions within a cell. Each type of reaction is catalyzed by a specific enzyme.

- A metabolic pathway is the sequence of chemical reactions needed to make a particular biological molecule. If a disruption occurs somewhere along the pathway, then the organism might develop an illness or suffer a deficiency.

IS THAT A FACT!

- When bears sleep in their dens during winter, their body temperature decreases several degrees. We know this because scientists crawled into the dens and took their temperatures. The lower body temperature reduces energy requirements, so bears can sleep for weeks or months without eating.

For background information about teaching strategies and issues, refer to the *Professional Reference for Teachers.*

Pre-Reading Questions

Students may not know the answers to these questions before reading the chapter, so accept any reasonable response.

Suggested Answers

1. All living things have cells, reproduce, sense and respond to change, have DNA, use energy, and grow and develop.

2. All organisms need food, water, air, and a place to live.

It's Alive!! Or, Is It?

 Pre-Reading Questions

1. What characteristics do all living things have in common?

2. What do organisms need in order to stay alive?

ROBOT BUGS!

What does it mean to say something is *alive*? Machines have some of the characteristics of living things, but they do not have all of them. This amazing robot insect can respond to changes in its environment. It can walk over obstacles. It can perform some tasks. But it is still not alive. How is it like and unlike the living insect pictured here? In this chapter, you'll learn about the characteristics that all living things share.

internet connect

HRW On-line Resources	sci**LINKS** NSTA	Smithsonian Institution	CNN**fyi**.com
go.hrw.com For worksheets and other teaching aids, visit the HRW Web site and type in the keyword: **HSTALV**	**www.scilinks.com** Use the *sci*LINKS numbers at the end of each chapter for additional resources on the **NSTA** Web site.	**www.si.edu/hrw** Visit the Smithsonian Institution Web site for related on-line resources.	**www.cnnfyi.com** Visit the CNN Web site for current events coverage and classroom resources.

START-UP Activity

LIGHTS ON!

Living things respond to change. In this activity, you will work with a partner to see how eyes react to changes in light.

Procedure

1. Observe a classmate's eyes in a room with normal light. Find the pupil, which is the black area in the center of the colored part of the eye, and note its size.

2. Have your partner keep both eyes open, and have him or her cover each one with a cupped hand. Wait about 1 minute.

3. Instruct your partner to pull away both hands quickly. Immediately look at your partner's pupils. Record what happens.

4. Now briefly shine a **flashlight** into your partner's eyes. In your ScienceLog, record how this affects the pupils.

   Caution: Do not use the sun as the source of the light.

5. Change places with your partner, and repeat steps 1–4 so that your partner can observe your eyes.

Analysis

6. How did your partner's eyes respond to changes in the level of light?

7. How did changes in the size of your pupils affect your vision? What does this tell you about why pupils change size?

3

START-UP Activity

LIGHTS ON!

MATERIALS
FOR EACH GROUP: • flashlight

Safety Caution

Students must not use the sun as a source of light.

Answers to START-UP Activity

3. The pupils were enlarged.

4. The pupils became smaller.

6. The pupils enlarged when there was little light and became smaller when there was a lot of light.

7. The larger the pupils were, the brighter things appeared. In a dark environment, pupils become larger, so more light can enter the eye. The surroundings appear brighter and can be more easily seen. In a bright environment, the pupils become smaller and less light enters the eye. Students may infer that pupils change size in response to the amount of light available to the eye. This response helps us see clearly when light conditions change.

Focus

Characteristics of Living Things

This section explains the characteristics that describe living things. Students will learn that living things have cells; that they sense and respond to stimuli; that they reproduce, have DNA, and use energy; and that they grow and develop. Based on this information, students will be able to identify things as either living or nonliving.

Bellringer

Display this question on the board or an overhead projector:

What are four living and four nonliving things that you interact with or see everyday? (Sample answer: living: family members, pets, house plants and trees, birds, insects; nonliving: clothes, books, automobile, furniture, radio, sidewalk)

Sheltered English

1) Motivate

DISCUSSION

Stimuli Ask students what they do when they go outside and it is cold. (They put on a jacket or go back inside, where it is warmer.)

Explain that feeling cold is a stimulus and that their reaction to the cold is a response. Ask students how people use technology to improve their ability to respond to environmental stimuli. (Examples are furnaces and wood stoves to heat buildings, air conditioners to cool buildings, sunglasses to shield eyes from bright sunlight, greenhouses to extend plant growing seasons.)

Terms to Learn

cell	sexual
stimulus	reproduction
homeostasis	DNA
asexual	heredity
reproduction	metabolism

What You'll Do

- List the characteristics of living things.
- Distinguish between asexual reproduction and sexual reproduction.
- Define and describe homeostasis.

Figure 1 *Each of these organisms is made of only one cell.*

Figure 2 *Trillions of cells make up these organisms.*

4

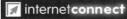

Directed Reading Worksheet 2 Section 1

internet connect

SCI LINKS
NSTA

TOPIC: Characteristics of Living Things
GO TO: www.scilinks.org
*sci*LINKS NUMBER: HSTL030

Characteristics of Living Things

While out in your yard one day, you notice something strange in the grass. It's slimy, bright yellow, and about the size of a dime. You have no idea what it is. Is it a plant part that fell from a tree? Is it alive? How can you tell?

Even though an amazing variety of living things exist on Earth, they are all alike in several ways. What does a dog have in common with a tree? What does a fish have in common with a mushroom? And what do *you* have in common with a slimy blob (also known as a slime mold)? Read on to find out about the six characteristics that all organisms share.

Slime mold

1 Living Things Have Cells

Every living thing is composed of one or more cells. A **cell** is a membrane-covered structure that contains all of the materials necessary for life. The membrane that surrounds a cell separates the contents of the cell from the cell's environment.

Many organisms, such as those in **Figure 1,** are made up of only one cell. Other organisms, such as the monkeys and trees in **Figure 2,** are made up of trillions of cells. Most cells are too small to be seen with the naked eye.

In an organism with many cells, cells perform specialized functions. For example, your nerve cells are specialized to transport signals, and your muscle cells are specialized for movement.

Science Bloopers

Leonardo da Vinci made many scientific discoveries and observations in the fifteenth and sixteenth centuries, but he mistakenly believed that the eye emitted a ray that struck and then rebounded from the observed object.

2 Living Things Sense and Respond to Change

All organisms have the ability to sense change in their environment and to respond to that change. When your pupils are exposed to light, they respond by becoming smaller. A change in an organism's environment that affects the activity of the organism is called a **stimulus** (plural, *stimuli*).

Stimuli can be chemicals, gravity, darkness, light, sounds, tastes, or anything that causes organisms to respond in some way. A gentle touch causes a response in the plant shown in **Figure 3**.

✔ **Self-Check**

Is your alarm clock a stimulus? Explain. *(See page 168 to check your answer.)*

Figure 3 *The touch of an insect triggers the Venus' flytrap to quickly close its leaves.*

Homeostasis Even though an organism's external environment may change, the organism must maintain a stable internal environment to survive. This is because the life processes of organisms involve many different kinds of chemical reactions that can occur only in delicately balanced environments. The maintenance of a stable internal environment is called **homeostasis** (нон mee oн STAY sis).

Your body maintains a temperature of about 37°C. When you get hot, your body responds by sweating. When you get cold, your muscles twitch in an attempt to generate heat. This causes you to shiver. Whether you are sweating or shivering, your body is trying to return things to normal. Another example of homeostasis is your body's ability to maintain a stable amount of sugar in your blood.

Oceanography
CONNECTION

Fish that live in the ice-cold waters off Antarctica make a natural antifreeze that keeps them from freezing.

WEIRD SCIENCE

The first indication that the pancreas was the organ that secreted insulin, the compound that removes sugar from the blood, came when flies were noticed swarming over the urine of a dog whose pancreas was damaged. The flies were attracted to the excess sugar in the urine.

Critical Thinking Worksheet
"Intergalactic Planetary Mission"

2 Teach

GUIDED PRACTICE

Writing Have students list stimuli that they experience in their lives. Write them on the board. (Sample answers: heat, cold, a red traffic light)

Ask each student to write this list on a sheet of paper and then to write how a person would respond to each stimulus. Sheltered English

Answer to Self-Check

Your alarm clock is a stimulus. It rings, and you respond by shutting it off and getting out of bed.

MEETING INDIVIDUAL NEEDS

 Learners Having Difficulty Have students use pictures from magazines to create a poster that shows three living and three non-living things. Underneath each picture, have students write a brief paragraph explaining the characteristics that identify each thing as either living or nonliving. Sheltered English

MATH and MORE

Red blood cells are very small. Healthy humans have about 5 million red blood cells per milliliter (mL) of blood. If there are 1,000 mL per liter, how many cells are there in a liter? (5 million × 1,000 = 5 billion, or 5,000,000,000)

 Math Skills Worksheet "A Shortcut for Multiplying Large Numbers"

 Math Skills Worksheet "Multiplying and Dividing in Scientific Notation"

5

DEBATE

Nature Versus Nurture

Scientists have proven that we inherit our physical characteristics from our parents (nature). They continue to research whether we inherit our personalities from our parents. Some scientists say that where we live and how we are raised are more important (nurture). What do students think is the critical factor, nurture (care) or nature (heredity)? Why?

MATH and MORE

The red kangaroo can cover 12 m in a single jump. The African sharp-nosed frog can leap 5.35 m. What percentage of the kangaroo's jump is the frog's leap?

(5.35 ÷ 12 = 0.4458, or 0.45
0.45 × 100 = 45%)

The common flea can leap 19 cm. What percentage of the frog's leap is the flea's?

(5.35 × 100 = 535 cm
19 ÷ 535 = 0.0355, or 0.04
0.04 × 100 = 4%)

 Math Skills Worksheet
"Decimals and Fractions"

 Math Skills Worksheet
"Percentages, Fractions, and Decimals"

Figure 4 *The hydra can reproduce asexually by forming buds that will break off and grow into new individuals.*

Figure 5 *Like most animals, bears produce offspring by sexual reproduction.*

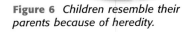

Figure 6 *Children resemble their parents because of heredity.*

3 Living Things Reproduce

Organisms make other organisms like themselves. This is accomplished in one of two ways: by asexual reproduction or by sexual reproduction. In **asexual reproduction,** a single parent produces offspring that are identical to the parent. **Figure 4** shows an organism that reproduces asexually. Most single-celled organisms reproduce in this way. **Sexual reproduction,** however, almost always requires two parents to produce offspring that will share characteristics of both parents. Most animals and plants reproduce in this way. The bear cubs in **Figure 5** were produced sexually by their parents.

4 Living Things Have DNA

The cells of all living things contain a special molecule called **DNA** (**d**eoxyribo**n**ucleic **a**cid). DNA provides instructions for making molecules called *proteins*. Proteins take part in almost all of the activities of an organism's cells. Proteins also determine many of an organism's characteristics. When organisms reproduce, they pass on copies of their DNA to their offspring. The transmission of characteristics from one generation to the next is called **heredity.** Offspring, such as the children in **Figure 6,** resemble their parents because of heredity.

5 Living Things Use Energy

Organisms use energy to carry out the activities of life. These activities include such things as making food, breaking down food, moving materials into and out of cells, and building cells. An organism's **metabolism** (muh TAB uh LIZ uhm) is the total of all of the chemical activities that it performs.

MISCONCEPTION ///**ALERT**\\\

Though very much alive, mules and most other hybrids cannot reproduce. Hybrids are the result of mating organisms from different species. A mule is the offspring of a mare (a female horse) and a jack (a male donkey). Mules often live long, healthy lives, but they never have babies.

IS THAT A FACT!

Lichens, which dominate the flora of Antarctica, have an extremely slow growth rate. Certain species grow only 1 mm every 100 years. Scientists estimate that some lichens may be more than 5,000 years old.

Are Computers Alive?
Computers can do all kinds of things, such as storing information and doing complex calculations. Some computers have even been programmed to learn, that is, to get better and faster at solving problems over time. Do you think computers could become so advanced that they should be considered alive? Why or why not?

6 Living Things Grow and Develop

All living things, whether they are made of one cell or many cells, grow during periods of their lives. Growth in single-celled organisms occurs as the cell gets larger. Organisms made of many cells grow mainly by increasing their number of cells.

In addition to getting larger, living things may develop and change as they grow. Just like the organisms in **Figure 7**, you will pass through different stages in your life as you develop into an adult.

Figure 7 *Over time, acorns develop into oak seedlings, which become oak trees.*

SECTION REVIEW

1. What characteristics of living things does a river have? Is a river alive?

2. What does the fur coat of a bear have to do with homeostasis?

3. How is reproduction related to heredity?

4. **Applying Concepts** What are some stimuli in your environment? How do you respond to these stimuli?

internet**connect**

*SCi*LINKS
NSTA

TOPIC: Characteristics of Living Things
GO TO: www.scilinks.org
*sci*LINKS NUMBER: HSTL030

4 Close

Answer to APPLY

Answers will vary. Students should include the six characteristics of living things in their answer and explain why a computer could or could not do those things and be considered alive.

Quiz

1. An apple tree is a living thing. Can it make oranges? Why or why not? (No; living things reproduce only themselves, not different living things.)

2. What is the difference between growth and development? (Growth is an increase in size. Development is the change of form of an organism.)

3. Name three activities of an organism that require energy. (Organisms need energy to make food, to break down food, to move materials into and out of cells, and to build cell parts.)

ALTERNATIVE ASSESSMENT

Writing Have students read a story of their choice and find five examples of stimuli and responses. Then have students write an explanation of why it is important to be able to respond to all of these stimuli.

▼ *Answers to Section Review*

1. A river has energy (it moves) and can grow larger (after rain, or when snow melts). But it is not alive because it is not made of cells, cannot respond to stimuli, has no DNA, and cannot reproduce.

2. Homeostasis is the maintenance of a stable internal environment. The fur coat of a bear helps it keep a stable body temperature.

3. Heredity is the passing of characteristics from parents to offspring. When organisms reproduce, offspring inherit copies of their parents' DNA.

4. Answers will vary. Examples can include such things as the way something tasted, smelled, felt, sounded, or looked. Responses to the stimuli should describe the action or effect the stimuli produced in the student.

Focus

The Simple Bare Necessities of Life

This section identifies the things that an organism needs to live. Students will learn the roles that food, water, and air play in an organism's survival. They will also learn that where an organism lives is related to its ability to obtain the necessities of life.

Bellringer

Pose the following question to students:

What do you think your mass would be if there were no water in your body? Write your answer in your ScienceLog. (The human body is approximately 70 percent water. If a student has a mass of 40 kg, the water's mass is 40 kg × 0.70 = 28 kg. The student's mass without water would be 40 kg − 28 kg = 12 kg.)

1 Motivate

DISCUSSION

Adaptation Show students pictures of a desert, the canopy of a rain forest, the Arctic, and seaside cliffs. Ask students to describe animals that could live in these places and to explain how each is adapted to obtain its necessities.

Directed Reading Worksheet Section 2

Terms to Learn

producer
consumer
decomposer

What You'll Do

◆ Explain why organisms need food, water, air, and living space.
◆ Discuss how living things obtain what they need to live.

Figure 8 *The salamander is a consumer. The fungus is a decomposer, and the plants are producers.*

The Simple Bare Necessities of Life

Would it surprise you to learn that you have the same basic needs as a tree, a frog, or a fly? In fact, almost every organism has the same basic needs: food, water, air, and living space.

Food

All living things need food. Food provides organisms with the energy and raw materials needed to carry on life processes and to build and repair cells and body parts. But not all organisms get food in the same way. In fact, organisms can be grouped into three different categories based on how they get their food.

Making Food Some organisms, such as plants, are called **producers** because they can produce their own food. Like most producers, plants use energy from the sun to make food from water and carbon dioxide. Some producers, like the microorganisms in Movile Cave, obtain energy and food from the chemicals in their environment.

Getting Food Other organisms are called **consumers** because they must eat (consume) other organisms to get food. The salamander in **Figure 8** is an example of a consumer. It gets the energy it needs by eating insects and other organisms.

Some consumers are decomposers. **Decomposers** are organisms that get their food by breaking down the nutrients in dead organisms or animal wastes.

Water

You may have heard that your body is made mostly of water. In fact, your cells and the cells of almost all living organisms are approximately 70 percent water—even the cells of a cactus and a camel. Most of the chemical reactions involved in metabolism require water.

Organisms differ greatly in terms of how much water they need and how they obtain it. You could survive for only about 3 days without water. You obtain water from the fluids you drink and the food you eat. The desert-dwelling kangaroo rat never drinks. It gets all of its water from its food.

CONNECT TO EARTH SCIENCE

Use Teaching Transparency 178 to encourage student discussion of how animals thrive in so many different climates and biomes.

Teaching Transparency 178 "Climate Zones of the Earth"
LINK TO EARTH SCIENCE

---

*internet***connect**

SCiLINKS
NSTA

TOPIC: The Necessities of Life
GO TO: www.scilinks.org
*sci***LINKS NUMBER:** HSTL035

Air

Air is a mixture of several different gases, including oxygen and carbon dioxide. Animals, plants, and most other living things use oxygen in the chemical process that releases energy from food. Organisms that live on land get oxygen from the air. Organisms living in fresh water and salt water either take in dissolved oxygen from the water or come to the water's surface to get oxygen from the air. Some organisms, such as the European diving spider in **Figure 9,** go to great lengths to get oxygen.

Green plants, algae, and some bacteria need carbon dioxide gas in addition to oxygen. The food these organisms produce is made from carbon dioxide and water by *photosynthesis* (FOHT oh SIN thuh sis), the process that converts the energy in sunlight to energy stored in food.

Figure 9 *This spider surrounds itself with an air bubble so that it can obtain oxygen underwater.*

A Place to Live

All organisms must have somewhere to live that contains all of the things they need to survive. Some organisms, such as elephants, require a large amount of space. Other organisms, such as bacteria, may live their entire life in a single pore on the tip of your nose.

Because the amount of space on Earth is limited, organisms often compete with each other for food, water, and other necessities. Many animals, including the warbler in **Figure 10,** will claim a particular space and try to keep other animals away. Plants also compete with each other for living space and for access to water and sunlight.

Figure 10 *A warbler's song is more than just a pretty tune. The warbler is protecting its home by telling other warblers to stay out of its territory.*

SECTION REVIEW

1. Why are decomposers categorized as consumers? How do they differ from producers?

2. Why are most cells 70 percent water?

3. **Making Inferences** Could life as we know it exist on Earth if air contained only oxygen? Explain.

4. **Identifying Relationships** How might a cave, an ant, and a lake meet the needs of an organism?

internet connect

SCi LINKS.
NSTA

TOPIC: The Necessities of Life
GO TO: www.scilinks.org
*sci*LINKS NUMBER: HSTL035

▼ *Answers to Section Review*

1. Decomposers are consumers because they must obtain the food they need from other organisms. Unlike producers, decomposers cannot produce their own food.

2. Most of the chemical reactions that occur in cells depend on the presence of water.

3. Life could not exist as we know it. Green plants, algae, and some bacteria need

carbon dioxide gas as well as oxygen. Without the carbon dioxide, they could not survive, and other organisms could not rely on them as a food source.

4. Answers will vary. A cave could be a place to live. An ant could be food. A lake could be a place to live as well as a source of water.

---

2) Teach

DEMONSTRATION

Fire and Life Demonstrate that both a human and a burning candle share some qualities of life. Briefly hold a cold drinking glass that is inverted over a candle flame. The glass will be fogged with water droplets. Now, breathe into a cold glass. The same will happen. Besides giving off water, both use oxygen and fuel (food or wax) and give off carbon dioxide and energy.
Sheltered English

3) Close

Quiz

1. Give an example of a producer, consumer, and decomposer. (producer: any plant; consumer: any animal; decomposer: fungi)

2. What factors affect where a plant or animal lives? (competition with other organisms and the availability of water and food that is sufficient to meet the organism's needs)

ALTERNATIVE ASSESSMENT

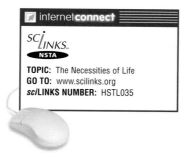

Writing Have students collect pictures from magazines and create a poster that shows the home of a plant and an animal with all the necessities of life that were discussed in this section. Then have students write a script for a nature documentary to accompany the poster.
Sheltered English

Reinforcement Worksheet
"Amazing Discovery"

Focus

The Chemistry of Life

In this section, students will learn about life on a cellular level. They will learn about proteins, carbohydrates, lipids, phospholipids, nucleic acids, and ATP and why they are essential to sustain life. Students will learn how these substances are alike and how they are different in both form and function.

🔊 Bellringer

Have students unscramble the following words and then use all four of them in a single sentence.

cdporesru (producers)
gnreey (energy)
dofo (food)
rwtea (water)

(Sample sentence: Producers use energy from the sun to make food from carbon dioxide and water.)

1) Motivate

DEMONSTRATION

Protein Model On each of 10–15 small self-adhesive notes, write a single letter of the alphabet. The letters should spell a few simple words when placed side by side. Place the note papers on two or three of the colored papers. Tell students that the binder contents represent proteins. The letters are the "amino acids." Each word is a "protein." Dismantle the "proteins," and use some of the "amino acids" to assemble a new "protein." **Sheltered English**

Terms to Learn

protein	phospholipid
carbohydrate	nucleic acid
lipid	ATP

What You'll Do

- ◆ Compare and contrast the chemical building blocks of cells.
- ◆ Explain the importance of ATP.

The Chemistry of Life

All living things are made of cells, but what are cells made of? Everything, whether it is living or not, is made up of tiny building blocks called *atoms*. There are about 100 different kinds of atoms.

A substance made up of one type of atom is called an *element*. When two or more atoms join together, they form what's called a *molecule*. Molecules found in living things are usually made of different combinations of six elements: carbon, hydrogen, nitrogen, oxygen, phosphorous, and sulfur. These elements combine to form proteins, carbohydrates, lipids, nucleic acids, and ATP.

Proteins

Almost all of the life processes of a cell involve proteins. After water, proteins are the most abundant materials in cells. **Proteins** are large molecules that are made up of subunits called *amino acids.*

Organisms break down the proteins in food to supply their cells with amino acids. These amino acids are then linked together to form new proteins. Some proteins are made up of only a few amino acids, while others contain more than 10,000 amino acids.

Proteins in Action Proteins have many different functions. Some proteins form structures that are easy to see, such as those in **Figure 11.** Other proteins are at work at the cellular level. The protein *hemoglobin* (HEE moh GLOH bin) in red blood cells attaches to oxygen so that oxygen can be delivered throughout the body. Some proteins help protect cells from foreign materials. And special proteins called *enzymes* make many different chemical reactions in a cell occur quickly.

Figure 11 *Feathers, spider webs, and hair are all made of proteins.*

🌐 Multicultural CONNECTION

Hunters and gatherers of all races historically required very large areas of land to sustain them. Most cultures later developed farming and herding techniques that made possible higher population densities. Staple crops vary around the world, but the millet and sorghum grains of Africans, the wheat and barley of Europeans, the corn and squash of Native Americans, and the rice and soybeans of Asians, in correct amounts and supplemented with other foods, all are equally nutritious.

Carbohydrates

Carbohydrates are a group of compounds made of sugars. Cells use carbohydrates as a source of energy and for energy storage. When an organism needs energy, its cells break down carbohydrates to release the energy stored in the carbohydrates.

There are two types of carbohydrates, simple carbohydrates and complex carbohydrates. Simple carbohydrates are made of one sugar molecule or a few sugar molecules linked together. Table sugar and the sugar in fruits are examples of simple carbohydrates.

Too Much Sugar! When an organism has more sugar than it needs, its extra sugar may be stored in the form of complex carbohydrates. Complex carbohydrates are made of hundreds of sugar molecules linked together. Your body makes some complex carbohydrates and stores them in your liver. Plants make a complex carbohydrate called *starch*. A potato plant, such as the one in **Figure 12,** stores its extra sugar as starch. When you eat mashed potatoes or French fries, you are eating a potato plant's stored starch. Your body can then break down this complex carbohydrate to release the energy stored in it.

Sugars

Starch

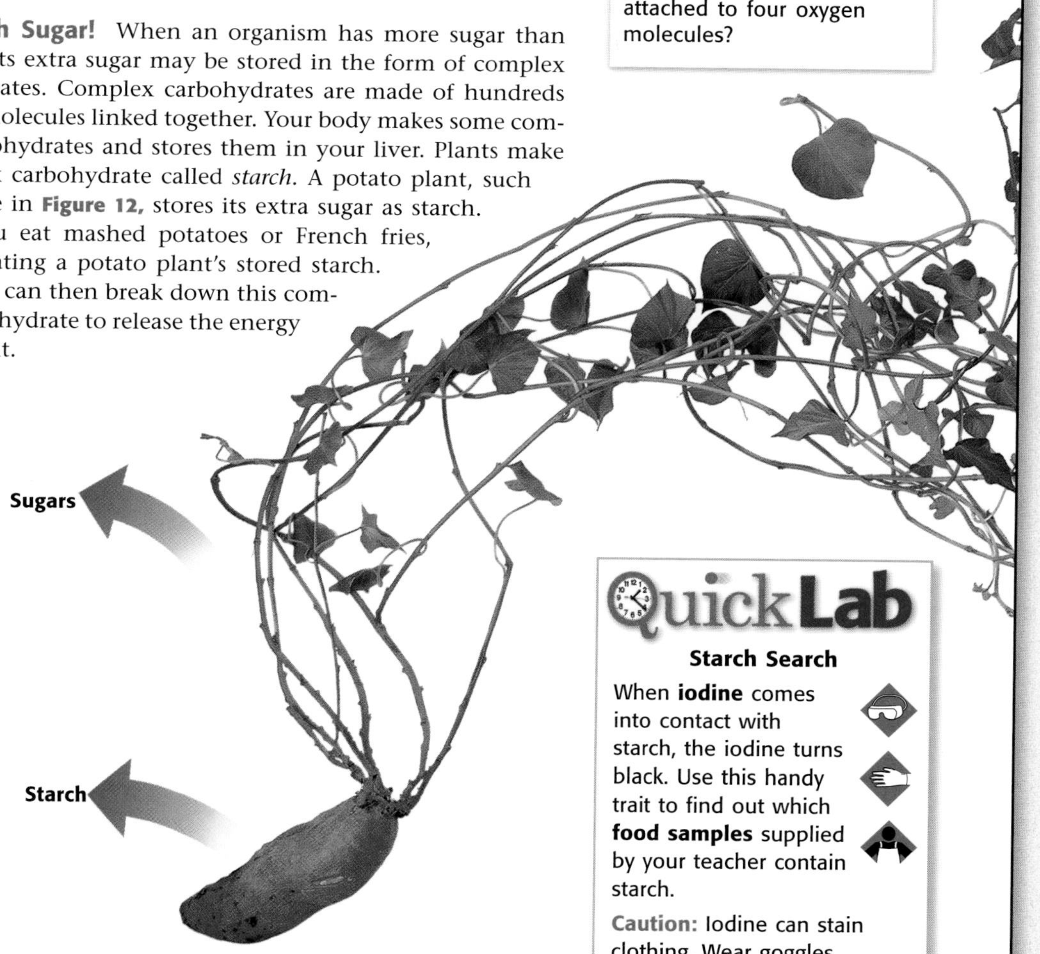

Figure 12 *Most sugars are simple carbohydrates. The extra sugar in a potato plant is stored in the potato as starch, a complex carbohydrate.*

MATH BREAK

How Much Oxygen?

Each red blood cell carries about 250 million molecules of hemoglobin. How many molecules of oxygen could a single red blood cell deliver throughout the body if every hemoglobin molecule attached to four oxygen molecules?

QuickLab

Starch Search

When **iodine** comes into contact with starch, the iodine turns black. Use this handy trait to find out which **food samples** supplied by your teacher contain starch.

Caution: Iodine can stain clothing. Wear goggles, protective gloves, and an apron.

11

WEIRD SCIENCE

Fireflies produce their flashing light by a chemical reaction. The enzyme luciferase acts on the chemical luciferin in the presence of ATP to create the light. Scientists now use luciferase in the laboratory to study everything from heart disease to muscular dystrophy.

internet connect

SCILINKS
NSTA

TOPIC: The Chemistry of Life
GO TO: www.scilinks.org
sciLINKS NUMBER: HSTL040

2 Teach

Answer to MATHBREAK

1 billion molecules of oxygen

DISCUSSION

Food Choices Ask students which snack foods they prefer to eat before playing or participating in sports. Write their responses on the board. Ask students why they prefer these foods. Taste? Simply to eliminate hunger? Do they think these foods give them more energy? Explain that this section may change their opinions about the foods they eat.

QuickLab

MATERIALS

FOR EACH GROUP:
• iodine solution in small bottle
• plastic eyedropper
• 25 × 25 cm piece of aluminum foil to hold food samples
• cracker, small piece of bread, potato, chocolate, apple, broccoli, celery, piece of hot dog

Safety Caution: Remind students to review all safety cautions and icons before beginning this lab activity.

Students should not eat any of the food samples. Iodine will stain and can be toxic. Dilute the iodine to prevent injury. Have a functioning eyewash available. Each student should wear safety goggles, an apron, and protective gloves.

Instruct students to use only a few drops of iodine on each sample.

Directed Reading Worksheet Section 3

③ Extend

LabBook **PG 134**

The Best-Bread Bakery Dilemma

MEETING INDIVIDUAL NEEDS

Advanced Learners

When the human body is unable to make a necessary protein, the result is often a disease. Hemophilia and diabetes are two conditions caused by missing or defective proteins. Have students prepare a report on the cause of and the treatment for one of these conditions. Their information should include the specific protein that is lacking, a brief description of how the condition affects the body, and the role of DNA in the disease.

Homework

Concept Mapping Have students collect the nutrition labels from five food items in their home and examine the number of grams of carbohydrates, fats, and proteins in each item. Tell students to construct a concept map that best relates the food items with the headings "Carbohydrates," "Lipids," and "Proteins," based on the nutrient content. If an item belongs in two categories, the map should reflect that information.

Teaching Transparency 5
"Phospholipid Molecule and Cell Membrane"

Lipids

Lipids are compounds that cannot mix with water. Lipids have many important functions in the cell. Like carbohydrates, some lipids store energy. Other lipids form the membranes of cells.

Fats and Oils Fats and oils are lipids that store energy. When an organism has used up most of its carbohydrates, it can obtain energy from these lipids. The structures of fats and oils are almost identical, but at room temperature most fats are solid and oils are liquid. Most of the lipids stored in plants are oils, while most of the lipids stored in animals are fats.

Phospholipids All cells are surrounded by a structure called a *cell membrane*. **Phospholipids** are the molecules that form much of the cell membrane. As you read earlier, water is the most abundant material in a cell. When phospholipids are in water, the tails come together and the heads face out into the water. This happens because the head of a phospholipid molecule is attracted to water, while the tail is not. **Figure 13** shows how phospholipid molecules form two layers when they are in water.

LabBook

Yeast cells get energy the same way other cells do. See for yourself on page 134 of the LabBook.

Figure 13 *The contents of a cell are surrounded by a membrane of phospholipid molecules.*

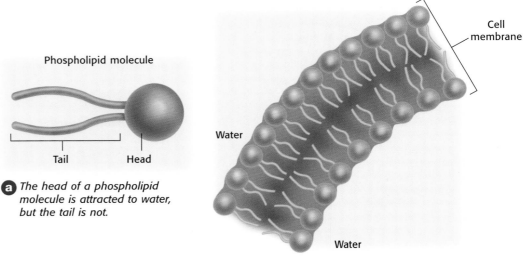

Phospholipid molecule

Tail   Head

ⓐ *The head of a phospholipid molecule is attracted to water, but the tail is not.*

Cell membrane

Water

Water

ⓑ *When phospholipid molecules come together in water, they form two layers.*

IS THAT A FACT!

Fats and lipids are also vital to nerve tissue formation and function. Fat acts as an insulator on the nerve, speeding up the electrical impulse that the nerve is carrying.

Nucleic Acids

Nucleic acids are compounds made up of subunits called *nucleotides*. A nucleic acid may contain thousands of nucleotides. Nucleic acids are sometimes called the blueprints of life because they contain all the information needed for the cell to make all of its proteins.

DNA is a nucleic acid. A DNA molecule is like a recipe book titled *How to Make Proteins*. When a cell needs to make a certain protein, it gets information from DNA to direct how amino acids are hooked together to make that protein. You will learn more about DNA later.

The Cell's Fuel

Another molecule that is important to cells is ATP (**a**denosine **trip**hosphate). **ATP** is the major fuel used for all cell activities that require energy.

When food molecules, such as carbohydrates and fats, are broken down, some of the released energy is transferred to ATP molecules, as shown in **Figure 14**. The energy in carbohydrates and lipids must be transferred to ATP before the stored energy can be used by cells to fuel their life processes.

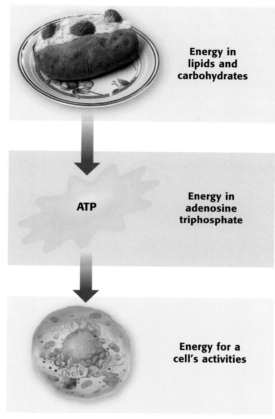

Energy in lipids and carbohydrates

ATP

Energy in adenosine triphosphate

Energy for a cell's activities

Figure 14 *The energy in the carbohydrates and lipids in food must be transferred to ATP molecules before cells can use the energy.*

SECTION REVIEW

1. What are the subunits of proteins? of starch? of DNA?

2. What do carbohydrates, fats, and oils have in common?

3. Are all proteins enzymes? Explain your answer.

4. **Making Predictions** What would happen to the supply of ATP in your cells if you did not eat enough carbohydrates? How would this affect your cells?

Quiz

1. Explain the difference between simple and complex carbohydrates. (Simple carbohydrates are made of one or two sugar molecules. Complex carbohydrates are made of many sugar molecules that are linked together.)

2. Name two functions of lipids. (Some lipids store energy, and others form the cell membrane.)

3. How are proteins used by an organism? (An organism breaks down proteins and uses their amino acids to build other proteins that are used to carry out chemical reactions in cells, transport materials, and protect the cell.)

ALTERNATIVE ASSESSMENT

Writing Have students write a job description for the cell's basic chemical building blocks. Tell students to write a classified ad that describes the required job responsibilities and physical qualifications. Have them include a description of the expected workload by explaining whether the building block will have to work constantly or sporadically. Finally, they should indicate whether the building block will work independently or with other cell components.

▼ Answers to Section Review

1. The subunits of proteins are amino acids. Sugar molecules are the subunits of starch, and nucleotides are the subunits of DNA.

2. All three compounds store energy.

3. Not all proteins are enzymes. Enzymes are a special type of protein that speeds up certain chemical reactions in the cell.

4. The supply of ATP would decrease. A decrease in ATP would cause a cell to have less energy than it needs to carry out its activities. Your body would have to get ATP from other sources, like lipids.

Teaching Transparency 6 "Energy for Cells"

Reinforcement Worksheet "Building Blocks"

Roly-Poly Races
Teacher's Notes

Time Required

One or two 45-minute class periods

Lab Ratings

EASY ———————→ HARD

TEACHER PREP   🧪🧪
STUDENT SET-UP 🧪🧪
CONCEPT LEVEL   🧪
CLEAN UP   🧪

MATERIALS

The materials listed on the student page are enough for 1–2 students. Remind students that they are handling living things that deserve to be treated with respect. The soil used in this lab should be sterilized potting soil to avoid causing allergic reactions among the students.

Safety Caution

Remind students to review all safety cautions and icons before beginning this lab activity.

Preparation Notes

Isopods were selected for this lab because they are very common in most areas and can be collected and released. If you choose to use other animals, such as mealworms, which you can obtain at a pet store, be sure to have a plan for appropriate disposal after the lab.

Roly-Poly Races

Have you ever watched a bug run? Did you wonder why it was running? The bug you saw running was probably reacting to a stimulus. In other words, something happened that made it run! One characteristic of living things is that they respond to stimuli. In this activity, you will study the movement of roly-polies. Roly-polies are also called pill bugs. They are not really bugs at all. They are land-dwelling crustaceans called isopods. Isopods live in dark, moist places, often under rocks or wood. You will provide stimuli to determine how fast your isopod can move and what affects its speed and direction. Remember that isopods are living things and must be treated gently and with respect.

MATERIALS

- small plastic container with lid
- 1 or 2 cm of soil for the container
- metric ruler
- small slice of raw potato
- piece of chalk
- 4 isopods
- watch or clock with a second hand

Procedure

1. Choose a partner and decide together how you will organize your roly-poly race. Discuss some gentle ways you might be able to stimulate your isopods to move. Choose five or six things that might cause movement, such as a gentle nudge or a change in temperature, sound, or light. Check your choices with your teacher.

2. In your ScienceLog or on a computer, make a data table similar to the one below. Label your columns with the stimuli you've chosen. Label the rows "Isopod 1," "Isopod 2," "Isopod 3," and "Isopod 4."

Isopod Responses			
	Stimulus 1: ?	Stimulus 2: ?	Stimulus 3: ?
Isopod 1			
Isopod 2			
Isopod 3			
Isopod 4			

14

Datasheets for LabBook

Science Skills Worksheet
"Doing a Lab Write-up"

CLASSROOM TESTED & APPROVED

Gladys Cherniak
St. Paul's Episcopal School
Mobile, Alabama

3 Place 1 or 2 cm of soil in a small plastic container. Add a small slice of potato and a piece of chalk. Your isopods will eat these things.

4 Place four isopods in your container. Observe them for a minute or two before you perform your tests. Record your observations in your ScienceLog.

5 Decide which stimulus you want to test first. Carefully arrange the isopods at the "starting line." The starting line can be an imaginary line at one end of the container.

6 Gently stimulate each isopod at the same time and in the same way. In your data table, record the isopods' responses to the stimulus. Be sure to measure and record the distance each isopod traveled. Don't forget to time the race.

7 Repeat steps 5–6 for each stimulus. Be sure to wait at least 2 minutes between trials.

Analysis

8 Describe the way isopods move. Do their legs move together?

9 Did your isopods move before or between the trials? Did the movement seem to have a purpose, or were the isopods responding to a stimulus? Explain your answer.

10 Did any of the stimuli you chose make the isopods move faster or go farther? Explain your answer.

Going Further
Isopods may not run for the joy of running like humans do. But humans, like all living things, react to stimuli. Describe three stimuli that might cause humans to run.

Answers
8. Isopods move their legs in sequence.
9. Answers will be based on students' observations and will vary.
10. Answers will vary.

Going Further
Answers will vary but might include the stimuli of fear, being late, joy, and a need for exercise.

15

Disposal Information
When you are done with the lab, dispose of the isopods according to your school's policy or release them into a natural area.

Chapter Highlights

Chapter Highlights

VOCABULARY DEFINITIONS

SECTION 1

cell a membrane-covered structure that contains all of the materials necessary for life

stimulus anything that affects the activity of an organism, organ, or tissue

homeostasis the maintenance of a stable internal environment

asexual reproduction reproduction in which a single parent produces offspring that are genetically identical to the parent

sexual reproduction reproduction in which two parents are required to produce offspring that will share characteristics of both parents

DNA deoxyribonucleic acid; hereditary material that controls all the activities of a cell, contains the information to make new cells, and provides instructions for making proteins

heredity the passing of traits from parent to offspring

metabolism the combined chemical processes that occur in a cell or living organism

SECTION 2

producer an organism that uses sunlight directly to make sugar

consumer an organism that eats producers or other organisms for energy

decomposer an organism that gets energy by breaking down the remains of dead organisms and consuming or absorbing the nutrients

---

SECTION 1

Vocabulary

cell (p. 4)
stimulus (p. 5)
homeostasis (p. 5)
asexual reproduction (p. 6)
sexual reproduction (p. 6)
DNA (p. 6)
heredity (p. 6)
metabolism (p. 6)

Section Notes

- All living things share the six characteristics of life.
- Organisms are made of one or more cells.
- Organisms detect and respond to stimuli in their environment.
- Organisms work to keep their internal environment stable so that the chemical activities of their cells are not disrupted. The maintenance of a stable internal environment is called homeostasis.
- Organisms reproduce and make more organisms like themselves. Offspring can be produced asexually or sexually.
- Offspring resemble their parents. The passing of characteristics from parent to offspring is called heredity.
- Organisms grow and may change during their lifetime.
- Organisms use energy to carry out the chemical activities of life. Metabolism is the sum of an organism's chemical activities.

SECTION 2

Vocabulary

producer (p. 8)
consumer (p. 8)
decomposer (p. 8)

Section Notes

- Organisms must have food. Producers make their own food. Consumers eat other organisms for food. Decomposers break down the nutrients in dead organisms and animal wastes.

☑ Skills Check

Math Concepts

HOW MANY? In the MathBreak on page 11, you determined how many molecules of oxygen a single red blood cell could carry.

$$\frac{250{,}000{,}000 \text{ molecules}}{1 \text{ red blood cell}} \times \frac{4 \text{ molecules of oxygen}}{1 \text{ molecule of hemoglobin}}$$

$$= 1{,}000{,}000{,}000 \text{ molecules of oxygen}$$

Visual Understanding

PHOSPHOLIPIDS Look at the illustrations of phospholipids and the cell membrane on page 12. Notice that the fluid inside and outside of the cell contains a lot of water. The head of the phospholipid molecule is attracted to water. Therefore, the phospholipid molecules that form the cell membrane line up with their tails facing away from the water.

16

Lab and Activity Highlights

Roly-Poly Races `PG 14`

The Best-Bread Bakery Dilemma `PG 134`

Datasheets for LabBook
(blackline masters for these labs)

SECTION 2

- Organisms depend on water. Water is necessary for maintaining metabolism.

- Organisms need oxygen to release the energy contained in their food. Plants, algae, and some bacteria also need carbon dioxide.

- Organisms must have a place to live where they can obtain the things they need.

SECTION 3

Vocabulary

protein (*p. 10*)
carbohydrate (*p. 11*)
lipid (*p. 12*)
phospholipid (*p. 12*)
nucleic acid (*p. 13*)
ATP (*p. 13*)

Section Notes

- Proteins, carbohydrates, lipids, nucleic acids, and ATP are important to life.

- Cells use carbohydrates for energy storage. Carbohydrates are made of sugars.

- Fats and oils store energy. Phospholipids make cell membranes.

- Proteins are made up of amino acids and have many important functions. Enzymes are proteins that help chemical reactions occur quickly.

- Nucleic acids are made up of nucleotides. DNA is a nucleic acid that contains the information for making proteins.

- Cells use molecules of ATP to fuel their activities.

Labs

The Best-Bread Bakery Dilemma (*p. 134*)

SECTION 3

protein a biochemical that is composed of amino acids; its functions include regulating chemical reactions, transporting and storing materials, and providing support

carbohydrate a biochemical composed of one or more simple sugars bonded together that is used as a source of energy and to store energy

lipid a type of biochemical that does not dissolve in water, including fats and oils; lipids store energy and make up cell membranes

phospholipid a molecule that forms much of a cell membrane

nucleic acid a biochemical that stores information needed to build proteins and other nucleic acids; made up of subunits called nucleotides

ATP adenosine triphosphate; molecule that provides energy for a cell's activities

---

 internet connect

GO TO: go.hrw.com

Visit the **HRW** Web site for a variety of learning tools related to this chapter. Just type in the keyword:

KEYWORD: HSTALV

SCI**LINKS**
N S T A
GO TO: www.scilinks.org

Visit the **National Science Teachers Association** on-line Web site for Internet resources related to this chapter. Just type in the *sci*LINKS number for more information about the topic:

TOPIC: Characteristics of Living Things	*sci*LINKS NUMBER: HSTL030
TOPIC: The Necessities of Life	*sci*LINKS NUMBER: HSTL035
TOPIC: The Chemistry of Life	*sci*LINKS NUMBER: HSTL040
TOPIC: Is There Life on Other Planets?	*sci*LINKS NUMBER: HSTL045

 Vocabulary Review Worksheet

 Blackline masters of these Chapter Highlights can be found in the **Study Guide**.

17

Lab and Activity Highlights

LabBank

 Labs You Can Eat, Say Cheese!

Long-Term Projects & Research Ideas,
I Think, Therefore I Live

USING VOCABULARY

1. homeostasis
2. heredity
3. consumer
4. carbohydrate/sugars
5. lipids

UNDERSTANDING CONCEPTS

Multiple Choice

6. d
7. b
8. a
9. b
10. c
11. c
12. a

Short Answer

13. Asexual reproduction can occur with just one parent, and offspring are identical to the parent. Two parents are usually required for sexual reproduction, and offspring share characteristics of both parents.

14. Living things must have air because both plants and animals need oxygen, which is one component of air. Producers also need carbon dioxide to make food.

15. ATP is the energy-containing molecule in a cell. It is the major fuel for all cellular activities.

Concept Mapping

16. An answer to this exercise can be found at the front of this book.

Concept Mapping Transparency 2

Blackline masters of this Chapter Review can be found in the **Study Guide.**

Chapter Review

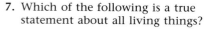

USING VOCABULARY

To complete the following sentences, choose the correct term from each pair of terms listed below:

1. The process of maintaining a stable internal environment is known as __?__. *(metabolism or homeostasis)*

2. The resemblance of offspring to their parents is a result of __?__. *(heredity or stimuli)*

3. A __?__ obtains food by eating other organisms. *(producer or consumer)*

4. Starch is a __?__ and is made up of __?__. *(carbohydrate/sugars or nucleic acid/nucleotides)*

5. Fats and oils are __?__ that store energy for an organism. *(proteins or lipids)*

UNDERSTANDING CONCEPTS

Multiple Choice

6. Cells are
   a. the structures that contain all of the materials necessary for life.
   b. found in all organisms.
   c. sometimes specialized for particular functions.
   d. All of the above

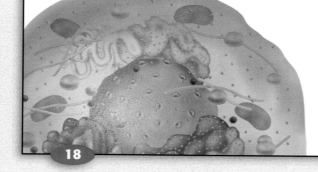

7. Which of the following is a true statement about all living things?
   a. They cannot sense changes in their external environment.
   b. They have one or more cells.
   c. They do not need to use energy.
   d. They reproduce asexually.

8. Organisms must have food because
   a. food is a source of energy.
   b. food supplies cells with oxygen.
   c. organisms never make their own food.
   d. All of the above

9. A change in an organism's environment that affects the organism's activities is a
   a. response.   c. metabolism.
   b. stimulus.   d. producer.

10. Organisms store energy in
    a. nucleic acids.   c. lipids.
    b. phospholipids.   d. water.

11. The molecule that contains the information on how to make proteins is
    a. ATP.
    b. a carbohydrate.
    c. DNA.
    d. a phospholipid.

12. The subunits of nucleic acids are
    a. nucleotides.   c. sugars.
    b. oils.   d. amino acids.

Short Answer

13. What is the difference between asexual reproduction and sexual reproduction?

14. In one or two sentences, explain why living things must have air.

15. What is ATP, and why is it important to a cell?

Concept Mapping

16. Use the following terms to create a concept map: cell, carbohydrates, protein, enzymes, DNA, sugars, lipids, nucleotides, amino acids, nucleic acid.

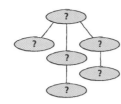

CRITICAL THINKING AND PROBLEM SOLVING

Write one or two sentences to answer the following questions:

17. A flame can move, grow larger, and give off heat. Is a flame alive? Explain.

18. Based on what you know about carbohydrates, lipids, and proteins, why is it important for you to eat a balanced diet?

19. Your friend tells you that the stimulus of music makes his goldfish swim faster. How would you design a controlled experiment to test your friend's claim?

MATH IN SCIENCE

20. An elephant has a mass of 3,900 kg. If 70 percent of the elephant's mass comes from water, how many kilograms of water does the elephant contain?

INTERPRETING GRAPHICS

Take a look at the pictures below, which show the same plant over a time span of 3 days.

Day 1

Day 2

Day 3

21. What is the plant doing?

22. What characteristic(s) of living things is the plant exhibiting?

Reading Check-up

Take a minute to review your answers to the Pre-Reading Questions found at the bottom of page 2. Have your answers changed? If necessary, revise your answers based on what you have learned since you began this chapter.

19

CRITICAL THINKING AND PROBLEM SOLVING

17. The flame is not alive. Although a flame can move, grow, and give off heat, it does not have all of the characteristics of an organism. For example, a flame is not made up of cells and does not contain DNA.

18. You get the raw materials to make the carbohydrates, lipids, and proteins that your body needs from the foods you eat. Carbohydrates and lipids provide energy for cells, and proteins provide amino acids to make the proteins your cells need. You need all three for the body to function efficiently.

19. A controlled experiment to test your friend's claim would consist of goldfish in a control group and goldfish in an experimental group. The control group would not be exposed to music, but the goldfish in the experimental group would be exposed to music. The goldfish in both groups would need to be of the same age and kind. They would have to be fed the same type and amount of food. The environments of the fish would have to be identical in such things as temperature and water quality.

MATH IN SCIENCE

20. The elephant contains 2,730 kg of water.

INTERPRETING GRAPHICS

21. The plant is bending toward the light coming through the window.

22. The plant is sensing a stimulus (the light) and responding to it.

Background

Scientists have always been intrigued by the possibility of life on other planets. Only recently have advanced biological and chemical techniques been sufficient to begin to investigate such claims. Science is not always black and white, and scientists often have heated debates over scientific evidence. Students should appreciate that scientists often have very different ways of interpreting the same evidence and that science does not always provide a clear-cut answer.

The scientists who question the studies of the ALH84001 meteorite argue that each component of the evidence can be accounted for by an inorganic explanation. Supporters argue that if all of the evidence is considered collectively, they point toward an organic explanation. One point of debate is whether Mars was cool enough to support life. Many scientists believe that Mars was much cooler at the time the meteorite left Mars. Although the Martian crust appears very dry, research has indicated that water, a necessity for life, was present in low concentrations. Scientists believe that the Martian atmosphere was much thicker at one time and that many of the atmospheric components necessary for life, including carbon dioxide and oxygen, were probably much more abundant then.

SCIENTIFICDEBATE

Life on Mars?

In late 1996 the headlines read, "Evidence of Life on Mars." What kind of life? Aliens similar to those that we see in sci-fi movies? Some creature completely unlike any we've seen before? Not quite, but the story behind the headlines is no less fascinating!

An Unusual Spaceship

In 1996, a group of researchers led by NASA scientists studied a 3.8-billion-year-old meteorite named ALH84001. These scientists agree that ALH84001 is a potato-sized piece of the planet Mars. They also agree that it fell to Earth about 13,000 years ago. It was discovered in Antarctica in 1984. And according to the NASA team, ALH84001 brought with it evidence that life once existed on Mars.

Life-Form Leftovers

On the surface of ALH84001, scientists found certain kinds of *organic molecules* (molecules containing carbon). These molecules are similar to those left behind when living things break down substances for food. And when these scientists examined the interior of the meteorite, they found the same organic

▲ *This scanning electron micrograph image of a tube-like structure found within meteorite ALH84001 is thought to be evidence of life on Mars.*

molecules throughout. Because these molecules were spread throughout the meteorite, scientists concluded the molecules were not contamination from Earth. The NASA team believes these organic leftovers are strong evidence that tiny organisms similar to bacteria lived, ate, and died on Mars millions of years ago.

Dirty Water or Star Dust

Many scientists disagree that ALH84001 contains evidence of Martian life. Some of them argue that the organic compounds are contaminants from Antarctic meltwater that seeped into the meteorite.

Others argue that the molecules were created by processes involving very high temperatures. These scientists think the compounds were formed during star formation and ended up on Mars when it became a planet. Other supporters of this theory believe that the compounds were created during the formation of rocks on Mars. In either case, they argue that no life-forms could exist at such high temperatures and that these compounds could not be the result of living things.

The Debate Continues

Scientists continue to debate the evidence of ALH84001. They are looking for evidence specific to biological life, such as proteins, nucleic acids, and cellular walls. Other scientists are looking to Mars itself for more evidence. Some hope to find underground water that might have supported life. Others hope to gather soil and rock samples that might hold evidence that Mars was once a living planet. Until scientists have more evidence, the debate will continue.

Think About It

▶ If you went to Mars, what kinds of evidence would you look for to prove that life once existed there? How could the discovery of nucleic or amino acids prove life existed on Mars?

20

Answer to Think About It

Answers will vary. The discovery of nucleic acids, amino acids, and cell walls would be strong evidence of life on Mars because they are components of living organisms.

internetconnect

SCILINKS
NSTA

TOPIC: Is There Life on Other Planets?
GO TO: www.scilinks.org
*sci*LINKS NUMBER: HSTL045

Science Fiction

"They're Made Out of Meat"

by Terry Bisson

Two space travelers millions of light-years from home are visiting an uncharted sector of the universe to find signs of life. Their mission is to contact, welcome, and log any and all beings in this quadrant of the universe. Once they discover a living being, they must find a way to communicate with it.

During their mission they encounter a life-form quite unlike anything they have ever seen before. These unusual beings can think and communicate. They have even built a few simple machines, so they aren't exactly pond scum.

Nevertheless, the explorers have very strong doubts about adding this new species to the list of known life-forms in the universe. The creatures are just too strange and, well, disgusting. They just don't fit on the list. Besides, with their limited abilities, it is unlikely they will make contact with any of the other life-forms that dwell elsewhere in the universe.

Perhaps it might be better if the explorers agreed to pretend that they never encountered these beings at all. But the travelers' official duty is to contact and welcome all life-forms, no matter how ugly they are or what they are made of. Can they bring themselves to perform their official duty? Will anyone believe their story if they do?

You'll find out by reading Terry Bisson's short story "They're Made Out of Meat." This story is in the *Holt Anthology of Science Fiction.*

21

Further Reading If students like Terry Bisson's style, suggest more of his stories to students. Some of his works include the following:

Bears Discover Fire and Other Stories, Tor Books, New York City, 1993

"10:07:24," *Absolute Magazine,* 1995

"First Fire," *Science Fiction Age,* Sept 1998

"The Player," *Fantasy and Science Fiction Magazine,* Oct/Nov 1997

SCIENCE FICTION

"They're Made Out of Meat"
by Terry Bisson

A remarkable and intelligent life-form has been discovered at the far reaches of the universe, but it's tough to get excited about this find . . .

Teaching Strategy

Reading Level This is a relatively short story that should not be difficult for the average student to read and comprehend.

Background

About the Author Terry Bisson has written everything from comic books to short stories, novels, plays, how-to articles about writing, and news editorials. He has taken the scripts of several popular movies and converted them to novels. Some of Bisson's works have appeared in digital-audio format on the World Wide Web. "They're Made Out of Meat" is just one of several stories featured in the SciFi Channel's *Seeing Ear Theater.* In 1991, Bisson's short story "Bears Discover Fire" received the highest honors possible for science fiction writers—both the Nebula Award and the Hugo Award.

In addition to being a writer, Bisson has been an automobile mechanic, an editor, a publisher's consultant, and a teacher. Bisson teaches writing at Clarion University, in Pennsylvania, and at the New School for Social Research, in New York City. He also maintains a personal Web site full of interesting information, works by guest writers, and excerpts from his novels and stories.

Chapter Organizer

CHAPTER ORGANIZATION	TIME MINUTES	OBJECTIVES	LABS, INVESTIGATIONS, AND DEMONSTRATIONS
Chapter Opener pp. 22–23	45	National Standards: SAI 1	**Start-Up Activity,** Our Constant Companions, p. 23
Section 1 Bacteria	90	▶ Describe the characteristics of a prokaryotic cell. ▶ Explain how bacteria reproduce. ▶ Compare and contrast eubacteria and archaebacteria. UCP 1, 3, 5, SAI 1, LS 1b, 1c, 2a, 3a, 3b, 4b	**QuickLab,** Spying on Spirilla, p. 26 **Demonstration,** Bacterial Shapes, p. 26 in ATE
Section 2 Bacteria's Role in the World	90	▶ Explain why life on Earth depends on bacteria. ▶ List five ways bacteria are useful to people. ▶ Describe why some bacteria are harmful to people. UCP 4, ST 1, SPSP 5, HNS 1, 3, LS 1f; Labs UCP 2, 3, 5, SAI 1, LS 1c	**Design Your Own Lab,** Aunt Flossie and the Intruder, p. 136 **Datasheets for LabBook,** Aunt Flossie and the Intruder **Ecolabs & Field Activities,** Ditch's Brew **Labs You Can Eat,** Bacterial Buddies **Inquiry Lab,** It's an Invasion! **Interactive Explorations CD-ROM,** Scope it Out! *A **Worksheet** is also available in the **Interactive Explorations Teacher's Guide.***
Section 3 Viruses	90	▶ Explain how viruses are similar to and different from living things. ▶ List the four major virus shapes. ▶ Describe the two kinds of viral reproduction. UCP 1, SAI 1, 2, SPSP 1, 4, 5, LS 1b, 1f, 2a; Labs UCP 2, 5	**Demonstration,** Flame Tulips, p. 33 in ATE **Making Models,** Viral Decorations, p. 36 **Datasheets for LabBook,** Viral Decorations **Long-Term Projects & Research Ideas,** Bacteria to the Rescue!

*See page **T23** for a complete correlation of this book with the*

NATIONAL SCIENCE EDUCATION STANDARDS.

TECHNOLOGY RESOURCES

 Guided Reading Audio CD
English or Spanish, Chapter 2

 One-Stop Planner CD-ROM with Test Generator

 Interactive Explorations CD-ROM
CD 1, Exploration 3, Scope It Out!

 CNN Science, Technology, & Society, Fingerprinting *E. coli,* Segment 15

 Science Discovery Videodisc
Science Sleuths: The Biogenic Picnic

CLASSROOM WORKSHEETS, TRANSPARENCIES, AND RESOURCES	SCIENCE INTEGRATION AND CONNECTIONS	REVIEW AND ASSESSMENT
Directed Reading Worksheet **Science Puzzlers, Twisters & Teasers**		
Directed Reading Worksheet, Section 1 **Transparency 37,** Bacterial Reproduction: Binary Fission **Transparency 38,** The Most Common Shapes of Bacteria **Reinforcement Worksheet,** Bacteria Bonanza	**MathBreak,** Airborne Organisms, p. 24 **Math and More,** p. 25 in ATE **Connect to Physical Science,** p. 25 in ATE	**Homework,** pp. 26, 27 in ATE **Self-Check,** p. 27 **Section Review,** p. 28 **Quiz,** p. 28 in ATE **Alternative Assessment,** p. 28 in ATE
Directed Reading Worksheet, Section 2 **Critical Thinking Worksheet,** Bacterial Blastoff	**Real-World Connection,** p. 30 in ATE **Multicultural Connection,** p. 30 in ATE **Apply,** p. 31	**Homework,** p. 31 in ATE **Section Review,** p. 32 **Quiz,** p. 32 in ATE **Alternative Assessment,** p. 32 in ATE
Directed Reading Worksheet, Section 3 **Transparency 107,** Gold Crystal Structure **Transparency 39,** The Basic Shapes of Viruses **Transparency 40,** The Lytic Cycle **Math Skills for Science Worksheet,** Multiplying Whole Numbers **Reinforcement Worksheet,** The Lytic Cycle	**MathBreak,** Sizing Up a Virus, p. 33 **Chemistry Connection,** p. 34 **Connect to Earth Science,** p. 34 in ATE **Science, Technology, and Society:** Edible Vaccines, p. 42 **Health Watch:** Helpful Viruses, p. 43	**Section Review,** p. 35 **Quiz,** p. 35 in ATE **Alternative Assessment,** p. 35 in ATE

 internet **connect**

 Holt, Rinehart and Winston On-line Resources
go.hrw.com

For worksheets and other teaching aids related to this chapter, visit the HRW Web site and type in the keyword: **HSTVIR**

 National Science Teachers Association
www.scilinks.org

Encourage students to use the *sci*LINKS numbers listed in the internet connect boxes to access information and resources on the **NSTA** Web site.

END-OF-CHAPTER REVIEW AND ASSESSMENT

Chapter Review in Study Guide
Vocabulary and Notes in Study Guide
Chapter Tests with Performance-Based Assessment, Chapter 2 Test
Chapter Tests with Performance-Based Assessment, Performance-Based Assessment 2
Concept Mapping Transparency 10

Chapter Resources & Worksheets

Visual Resources

TEACHING TRANSPARENCIES

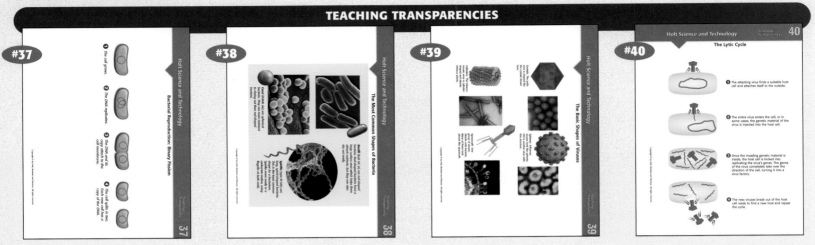

#37 — Bacterial Reproduction: Binary Fission

#38 — The Most Common Shapes of Bacteria

#39 — The Basic Shapes of Viruses

#40 — The Lytic Cycle

TEACHING TRANSPARENCIES

#107 — Gold Crystal Structure

LINK TO EARTH SCIENCE

CONCEPT MAPPING TRANSPARENCY

#10

Holt Science and Technology

Bacteria and Viruses

Use the following terms to complete the concept map below:
bacteria, archaebacteria, decomposers, eubacteria, prokaryotes, nucleus, consumers, producers

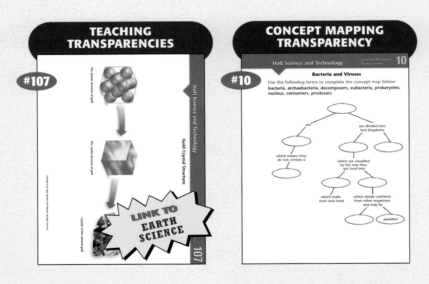

Meeting Individual Needs

DIRECTED READING

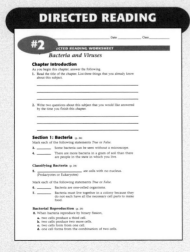

#2 — DIRECTED READING WORKSHEET

Bacteria and Viruses

Chapter Introduction

REINFORCEMENT & VOCABULARY REVIEW

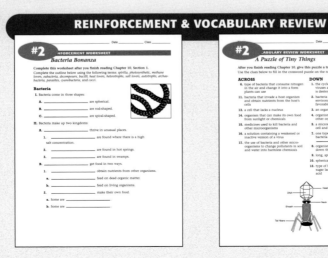

#2 — REINFORCEMENT WORKSHEET

Bacteria Bonanza

#2 — VOCABULARY REVIEW WORKSHEET

A Puzzle of Tiny Things

SCIENCE PUZZLERS, TWISTERS & TEASERS

#2 — SCIENCE PUZZLERS, TWISTERS & TEASERS

Bacteria and Viruses

Review & Assessment

STUDY GUIDE

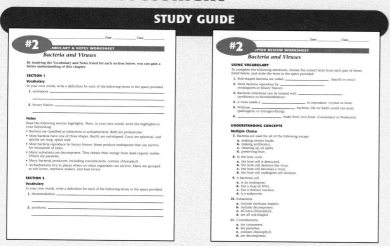

#2 — VOCABULARY & NOTES WORKSHEET
Bacteria and Viruses

By studying the Vocabulary and Notes listed for each section below, you can gain a better understanding of this chapter.

SECTION 1
Vocabulary
In your own words, write a definition for each of the following terms in the space provided.
1. endospore
2. binary fission

Notes
Read the following section highlights. Then, in your own words, write the highlights in your ScienceLog.
• Bacteria are classified as eubacteria or archaebacteria. Both are prokaryotes.
• Most bacteria have one of three shapes. Bacilli are rod-shaped. Cocci are spherical, and spirilla are long, spiral rods.
• Most bacteria reproduce by binary fission. Some produce endospores that can survive for thousands of years.
• Many eubacteria are decomposers. They obtain their energy from dead organic matter. Others are parasites.
• Many bacterial producers, including cyanobacteria, contain chlorophyll.
• Archaebacteria live in places where no other organisms can survive. Many are grouped as salt lovers, methane makers, and heat lovers.

SECTION 2
Vocabulary
In your own words, write a definition for each of the following terms in the space provided.
1. bioremediation
2. antibiotic

#2 — CHAPTER REVIEW WORKSHEET
Bacteria and Viruses

USING VOCABULARY
To complete the following sentences, choose the correct term from each pair of terms listed below, and write the term in the space provided.
1. Rod-shaped bacteria are called _____ (bacilli or cocci)
2. Most bacteria reproduce by _____ (endospores or binary fission)
3. Bacterial infections can be treated with _____ (antibiotics or bioremediation)
4. A virus needs a _____ to reproduce. (crystal or host)
5. Without _____ bacteria, life on Earth could not exist. (pathogenic or nitrogen-fixing)
6. _____ make their own food. (Consumers or Producers)

UNDERSTANDING CONCEPTS
Multiple Choice
7. Bacteria are used for all of the following except
   a. making certain foods.
   b. making antibiotics.
   c. cleaning up oil spills.
   d. preserving fruit.
8. In the lytic cycle
   a. the host cell is destroyed.
   b. the host cell destroys the virus.
   c. the host cell becomes a virus.
   d. the host cell undergoes cell division.
9. A bacterial cell
   a. is an endospore.
   b. has a loop of DNA.
   c. has a distinct nucleus.
   d. is a eukaryote.
10. Eubacteria
   a. include methane makers.
   b. include decomposers.
   c. all have chlorophyll.
   d. are all rod-shaped.
11. Cyanobacteria
   a. are consumers.
   b. are parasites.
   c. contain chlorophyll.
   d. are decomposers.

CHAPTER TESTS WITH PERFORMANCE-BASED ASSESSMENT

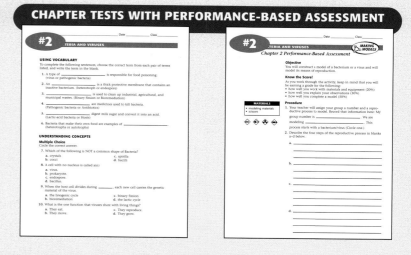

#2 — BACTERIA AND VIRUSES

USING VOCABULARY
To complete the following sentences, choose the correct term from each pair of terms listed, and write the term in the blank.
1. A type of _____ is responsible for food poisoning. (virus or pathogenic bacteria)
2. An _____ is a thick protective membrane that contains an inactive bacterium. (heterotroph or endospore)
3. _____ is used to clean up industrial, agricultural, and municipal wastes. (Pathogenic bacteria or Bioremediation)
4. _____ are medicines used to kill bacteria. (Pathogenic bacteria or Antibiotics)
5. _____ digest milk sugar and convert it into an acid. (Lactic-acid bacteria or Hosts)
6. Bacteria that make their own food are examples of _____ (heterotrophs or autotrophs)

UNDERSTANDING CONCEPTS
Multiple Choice
Circle the correct answer.
7. Which of the following is NOT a common shape of bacteria?
   a. crystals   c. spirilla
   b. cocci   d. bacilli
8. A cell with no nucleus is called a(n)
   a. virus.
   b. prokaryote.
   c. endospore.
   d. bacillus.
9. When the host cell divides during _____, each new cell carries the genetic material of the virus.
   a. the lysogenic cycle   c. binary fission
   b. bioremediation   d. the lactic cycle
10. What is the one function that viruses share with living things?
   a. They eat.   c. They reproduce.
   b. They move.   d. They grow.

#2 — BACTERIA AND VIRUSES — MAKING MODELS
Chapter 2 Performance-Based Assessment

Objective
You will construct a model of a bacterium or a virus and will model its means of reproduction.

Know the Score!
As you work through the activity, keep in mind that you will be earning a grade for the following.
• how well you work with materials and equipment (20%)
• how well you explain your observations (30%)
• how well you complete a model (50%)

MATERIALS
• modeling materials
• scissors

Procedure
1. Your teacher will assign your group a number and a reproductive process to model. Record that information here: My group number is _____. We are modeling _____. This process starts with a bacterium/virus (Circle one.)
2. Describe the four steps of the reproductive process in blanks a–d below.
   a.
   b.
   c.
   d.

Lab Worksheets

INQUIRY LABS

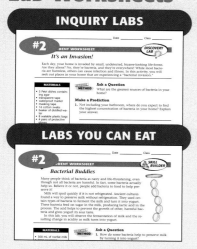

#2 — STUDENT WORKSHEET — DISCOVERY LAB
It's an Invasion!

Each day, your home is invaded by small, undetected, bizarre-looking life-forms. Are they aliens? No, they're bacteria, and they're everywhere! While most bacteria are harmless, others can cause infection and illness. In this activity, you will seek out places in your home that are experiencing a "bacterial invasion."

MATERIALS
• 5 Petri dishes containing agar
• transparent tape
• waterproof marker
• masking tape
• 16 cotton swabs
• beaker of distilled water
• 8 sealable plastic bags
• 4 pairs of protective gloves

Ask a Question
What are the greatest sources of bacteria in your home?

Make a Prediction
1. Not including your bathroom, where do you expect to find the highest concentration of bacteria in your home? Explain your answers.

LABS YOU CAN EAT

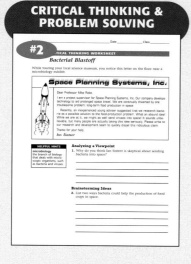

#2 — STUDENT WORKSHEET — SKILL BUILDER
Bacterial Buddies

Many people think of bacteria as nasty and life-threatening, even though not all bacteria are harmful. In fact, some bacteria actually help us. Believe it or not, people add bacteria to food to help preserve it!

Milk will spoil quickly if it is not refrigerated. Ancient cultures found a way to preserve milk without refrigeration. They used certain types of bacteria to ferment the milk and turn it into yogurt. These bacteria feed on sugar in the milk, producing lactic acid in the process. The acid helps to prevent the growth of other, harmful bacteria and gives yogurt its sour taste.

In this lab, you will observe the fermentation of milk and the resulting change in acidity as milk turns into yogurt.

MATERIALS
• 500 mL of nonfat milk

Ask a Question
1. How do some bacteria help to preserve milk by turning it into yogurt?

ECOLABS & FIELD ACTIVITIES

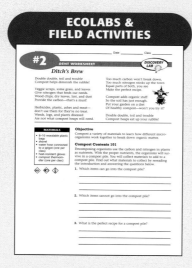

#2 — STUDENT WORKSHEET — DISCOVERY LAB
Ditch's Brew

Double double, toil and trouble
Compost helps diminish the rubble!

Veggie scraps, some grass, and leaves
Give nitrogen that feeds our needs.
Wood chips, dry leaves, lint, and dust
Provide the carbon—that's a must!

Herbicides, plastic, ashes and meat—
don't use them for they're no treat.
Weeds, logs, and plants diseased
Are not what compost heaps will need.

Too much carbon won't break down,
Too much nitrogen stinks up the town
Equal parts of both, you see
Make the perfect recipe.

Compost adds organic stuff
So the soil has just enough.
Put your garden on a diet
Of healthy compost—won't you try it?

Double double, toil and trouble
Compost heaps eat up your rubble!

MATERIALS
• 8–10 resealable plastic bags
• shovel
• water hose connected to a spigot (one per class)
• heat-resistant gloves
• compost thermometer (one per class)

Objective
Compost a variety of materials to learn how different microorganisms work together to break down organic matter.

Compost Contents 101
Decomposing organisms use the carbon and nitrogen in plants as nutrients. With the proper nutrients, the organisms will survive in a compost pile. You will collect materials to add to a compost pile. Find out what materials to collect by rereading the introduction and answering the questions below.
1. Which items can go into the compost pile?
2. Which items cannot go into the compost pile?
3. What is the perfect recipe for a compost pile?

LONG-TERM PROJECTS & RESEARCH IDEAS

#2 — STUDENT WORKSHEET — DESIGN YOUR OWN
Bacteria to the Rescue!

Imagine the aftermath of a supertanker accident. The clear blue ocean turns a murky black. The sea blazes with flames. Millions of liters of spilled oil threaten the lives of fish, birds, and other sea life. But microscopic organisms—bacteria—may be able to save them. Believe it or not, some bacteria eat oil like it was ice cream. Genera like *Pseudomonas* and *Penicillium* love oil spills and can actually make them disappear. Using bacteria to clean up pollution is called *bioremediation*.

INTERNET KEYWORDS
bioremediation
oil spill cleanup
ocean pollution

An Oily Yeast
1. How does bioremediation work best? In what situations does bioremediation work best? On which major oil spills has it been used? Is it more or less effective than more traditional cleanup methods? Are there disadvantages to using bioremediation? Ask your teacher to obtain cultures of bacteria that are used to clean up oil spills, or to get an oil spill kit from a scientific supply house. Create a model oil spill and use the bacteria to clean the oil from the water. Summarize the effectiveness of this form of bioremediation in a paper, stating your criteria and supporting your claims.

Other Long-Term Project Ideas
2. Are you being exposed to more germs than you should be? Visit a restaurant and either a hospital or biology lab, and compare the procedures used for sanitation at each. Ask a staff member to demonstrate the steps taken to prevent the spread of bacteria and viruses. How are the sanitation standards different? Do you think that each institution's procedures are thorough enough? Prepare a display board that shows what you have learned.
3. Your drinking water may have been through bioremediation! Visit a water treatment plant that uses bioremediation in its processing. Interview the water treatment manager. Tour the plant, taking pictures or videotaping during your tour, and prepare a presentation or documentary to share your discoveries with the class.

Research Idea
4. The bacterial and viral diseases that Europeans brought to the Americas had never been exposed to these diseases, and therefore, they had no resistance to them. Write a historical account of the devastation of an American Indian community as a result of contact with foreign diseases.

DATASHEETS FOR LABBOOK

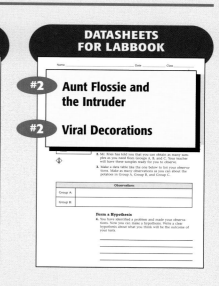

Name _____   Class _____

#2 **Aunt Flossie and the Intruder**

#2 **Viral Decorations**

2. Mr. Fries has told you that you can obtain as many samples as you need from Groups A, B, and C. Your teacher will have these samples ready for you to observe.
3. Make a data table like the one below to list your observations. Make as many observations as you can about the potatoes in Group A, Group B, and Group C.

Observations

| Group A: |
| Group B: |

Form a Hypothesis
4. You have identified a problem and made your observations. Now you can make a hypothesis. Write a clear hypothesis about what you think will be the outcome of your tests.

Applications & Extensions

CRITICAL THINKING & PROBLEM SOLVING

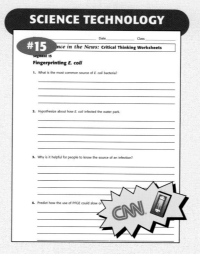

#2 — CRITICAL THINKING WORKSHEET
Bacterial Blastoff

While touring your local science museum, you notice this letter on the floor near a microbiology exhibit.

Space Planning Systems, Inc.

Dear Professor Mike Robe,

I am a project supervisor for Space Planning Systems, Inc. Our company develops technology to aid prolonged space travel. We are continually thwarted by one troublesome problem: long-term food production in space.

Recently, an inexperienced young adviser suggested that we research bacteria as a possible solution to the food-production problem. What an absurd idea! While we are at it, we might as well send viruses into space! It sounds unbelievable, but many people are actually taking this idea seriously. Please write to our research and development team to quickly dispel this ridiculous claim.

Thanks for your help,
Ian Sistent

HELPFUL HINTS
microbiology
the branch of biology that deals with microscopic organisms, such as bacteria and viruses

Analyzing a Viewpoint
1. Why do you think Ian Sistent is skeptical about sending bacteria into space?

Brainstorming Ideas
2. List two ways bacteria could help the production of food crops in space.

SCIENCE TECHNOLOGY

#15 — Science in the News: Critical Thinking Worksheets
Segment 15
Fingerprinting E. coli

1. What is the most common source of E. coli bacteria?
2. Hypothesize about how E. coli infected the water park.
3. Why is it helpful for people to know the source of an infection?
4. Predict how the use of PFGE could slow or

INTERACTIVE EXPLORATIONS

#1–3 — Exploration 3 Worksheet

Scope It Out!

1. What does Dr. Viola Ross need to know about the ancient microorganisms?
2. What does Dr. Ross intend to do with the results?
3. What will you use to conduct your investigation?
4. What do the ancient microorganisms look like under the microscope?

CD-ROM

SECTION 1

Bacteria

▶ Kingdoms of Bacteria

The classification of living things is an ongoing, ever-changing process. This book utilizes a six-kingdom system of classification in which bacteria are divided into two separate kingdoms: Eubacteria and Archaebacteria. In five-kingdom systems of classification, all bacteria are classified in the kingdom Monera.

- In addition to being classified by shape, many eubacteria also are classified as Gram-positive or Gram-negative. Gram-positive bacteria have a thick peptidoglycan cell wall that appears purple when stained with crystal violet, iodine, and safranine. Gram-negative bacteria have a much thinner peptidoglycan cell wall, as well as additional layers of lipids and polysaccharides that are not found in Gram-positive cells. Gram-negative bacteria appear red when stained with the same combination of dyes.

 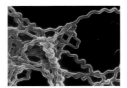

▶ Mistaken Identity

Cyanobacteria were once classified as blue-green algae because they sometimes grow together in long filaments that resemble algae and undergo photosynthesis. This classification was incorrect because cyanobacteria are prokaryotes and do not have nuclei or membrane-bound organelles. Blue-green algae, in contrast, are eukaryotes and have nuclei and membrane-bound organelles.

IS THAT A FACT!

- ➤ While some cyanobacteria are blue-green in color, cyanobacteria can also be yellow, red, brown, green, or black.

SECTION 2

Bacteria's Role in the World

▶ Bacteria and Plants

When the nitrogen-fixing bacteria *Rhizobium* enters a plant's roots, it forms a nodule. *Rhizobium* inhabits only the roots of legumes, which include beans, peas, soybeans, alfalfa, clover, peanuts, and vetch.

- One example of the effective biological control of insects involves the bacteria *Bacillus thunguriensis.* When sprayed or dusted onto a plant's leaves, the bacteria are an effective destroyer of leaf-eating insects, such as the tomato hornworm. When ingested by the insect, the bacteria begin to secrete enzymes that dissolve the insect's digestive system. Within 24 hours after ingestion, the insect stops eating and dies.

IS THAT A FACT!

- ➤ Bacteria are used in a number of industrial applications. They can peel and eat the paint off old aircraft. They can be used as biosensors to help determine the health of a specific habitat. Bacteria are even being used to remove the sulfur from coal before it is burned, which helps reduce acid rain.

▶ A Taste of Their Own Medicine

Common antibiotics produced by bacteria include streptomycin and tetracycline. A fungus produces penicillin, one of the most common antibiotics.

▶ Poison Producers

Botulism, a type of food poisoning, is caused by a toxin produced by the bacteria *Clostridium botulinum.* Consumption of very small amounts—sometimes as little as one-millionth of a gram—can cause paralysis and eventually death. *C. botulinum* grows in foods that have been improperly canned and sterilized.

- The intestinal disorder that many travelers refer to as traveler's diarrhea is frequently caused by a strain of the common intestinal bacteria *Escherichia coli,* commonly called *E. coli.* In many countries, contraction of the disease can be avoided by not drinking tap water and by not eating uncooked fruits and vegetables.

IS THAT A FACT!

➥ The water droplets in the air produced in a sneeze can carry between 10,000 and 100 million bacteria. The bacteria that cause whooping cough, tuberculosis, diphtheria, and scarlet fever can be carried through the air on droplets such as these.

SECTION 3

Viruses

▶ The Discovery of Viruses

Viruses were first discovered when scientists were trying to find the cause of tobacco mosaic disease, a disease in which the leaves of tobacco plants become wrinkled, blotchy, and yellow and have stunted growth.

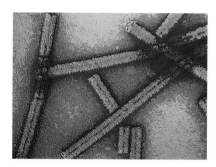

• In 1883, a German scientist named Adolf Mayer discovered that tobacco mosaic disease was contagious, and he suspected that "very small" bacteria caused the disease. Mayer's hypothesis was tested in 1892 by the Russian biologist Dmitri Ivanovsky, who passed the liquid from a diseased plant through extremely fine filters to isolate the small bacteria. He found no bacteria but discovered that the filtered sap from an infected plant caused the disease in a healthy plant. He concluded that "poison" from the bacteria was the cause of the disease. In 1897, a Dutch biologist named Martinus Beijernick discovered that the disease-causing agent was something smaller and simpler than bacteria. In 1898, Beijernick used the word *virus* to describe these tiny disease-causing agents.

• An American biochemist named Wendell Stanley (1904–1971) first isolated the tobacco mosaic virus in 1935. He treated the sap from infected plants in such a way that the viruses formed needle-shaped crystals. When Stanley spread the crystals from the diseased plant onto the leaves of a healthy plant, the healthy plant developed the disease.

• Even though Stanley isolated and chemically analyzed viruses in 1935, viruses were not seen until 1940, when the electron microscope was invented.

IS THAT A FACT!

➥ Most viruses are named for the disease they cause; for example, the cause of tobacco mosaic disease is tobacco mosaic virus.

➥ All viruses are parasitic, that is, they must have a host to reproduce. Because of this trait, viruses could not have been the first organisms on Earth.

▶ Vaccines

One of the greatest medical uses of vaccines was the elimination of smallpox infections from the world. Smallpox is a disease in which small pustules form on the victim's skin, causing permanent disfigurement and, in half the cases, death. In 1798, an English doctor named Edward Jenner (1749–1823) developed the first effective vaccine for smallpox. The use of smallpox vaccines rapidly spread in industrialized countries around the world. The last reported case in the United States occurred in 1949.

• Jonas Salk (1914–1995) developed a vaccine for the paralyzing disease polio in 1952.

IS THAT A FACT!

➥ A 23-year-old man named Ali Maow Maalin, of Merka, Somalia, contracted the last known case of smallpox in the world in 1977.

For background information about teaching strategies and issues, refer to the *Professional Reference for Teachers.*

CHAPTER 2

Bacteria and Viruses

 Pre-Reading Questions

Students may not know the answers to these questions before reading the chapter, so accept any reasonable response.

Suggested Answers

1. One cell makes up a bacterium.

2. Bacteria are used to make cheese and yogurt.

3. Sample answer: A virus is not alive because it cannot reproduce on its own.

Sections

 **Pre-Reading Questions**

1. How many cells make up a bacterium?

2. What do cheese and yogurt have to do with bacteria?

3. Do you think a virus is alive? Explain your answer.

22

BACTERIA: FRIEND AND FOE!!

Bacteria are everywhere. Some provide us with medicines and some make foods we eat. Others, like the one pictured here, can cause illnesses. This bacterium is a kind of *Salmonella,* and it can cause food poisoning. *Salmonella* can live inside chickens and other birds. Cooking eggs and chicken properly helps make sure that you don't get sick from *Salmonella.* In this chapter, you will learn more about bacteria and viruses.

internet connect

 HRW On-line Resources

go.hrw.com
For worksheets and other teaching aids, visit the HRW Web site and type in the keyword: **HSTVIR**

sci LINKS
NSTA

www.scilinks.com
Use the *sci*LINKS numbers at the end of each chapter for additional resources on the **NSTA** Web site.

 Smithsonian Institution®

www.si.edu/hrw
Visit the Smithsonian Institution Web site for related on-line resources.

 CNNfyi.com

www.cnnfyi.com
Visit the CNN Web site for current events coverage and classroom resources.

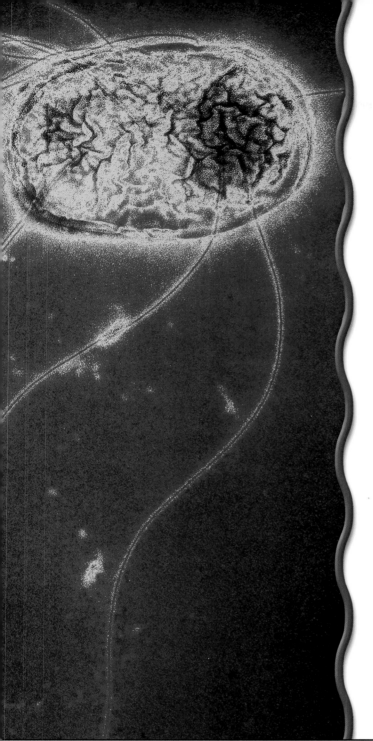

OUR CONSTANT COMPANIONS

Bacteria are everywhere. They are in the soil, in the air, and even inside you. When grown in a laboratory, microscopic bacteria form colonies that you can see. In this activity, you will see some of the bacteria that share your world.

Procedure

1. Get **three plastic Petri dishes** containing **nutrient agar** from your teacher. Label one dish "Hand," another "Breath," and another "Soil." Wipe your finger across the inside of the first dish. Breathe into the second dish. Place a small amount of **soil** into the third dish.

2. Secure the Petri dish lids with **transparent tape.** Wash your hands. Place the dishes in a warm, dark place for about 1 week. **Caution:** Do not open the Petri dishes after they are sealed.

3. Observe the Petri dishes each day. What do you see? Record your observations in your ScienceLog.

Analysis

4. How does the appearance of the agar in each dish differ?

5. Which source had the most bacterial growth—your hand, your breath, or the soil? Why do you think this might be?

23

OUR CONSTANT COMPANIONS

MATERIALS
FOR EACH GROUP:
• 3 plastic Petri dishes filled with nutrient agar
• soil
• transparent tape

Safety Caution

When students breathe on the dishes, they should be careful not to breathe on each other. Plastic dishes are safer than glass dishes.

Teacher's Notes

Prepoured agar plates can be purchased from a biological supply house.

To minimize the risk of contamination from airborne bacteria, students should only lift the Petri dish lids slightly when treating the agar.

Do not allow students to open the Petri dishes after they have been inoculated and sealed.

If moisture collects on the plates in step 2, try storing the plates upside down.

Answers to START-UP Activity

4. Answers will vary, but students should mention color differences, variations in colony shapes and sizes, and differing amounts of growth.

5. Answers will vary. Most likely, the Petri dish that is exposed to soil will have the most bacterial growth. All logical conclusions should be accepted.

Focus

Bacteria

This section introduces students to the characteristics of a prokaryotic cell. Students will compare and contrast eubacteria and archaebacteria. Students will distinguish between bacterial producers and consumers and will describe the three main groups of archaebacteria.

Bellringer

Pose the following question to your students, and have them write their answers in their ScienceLog:

If you can't see bacteria without a microscope, how do you know when you have come in contact with them? (Sample answer: Bacteria sometimes cause infected sores, and sometimes cause food to spoil.)

1) Motivate

DISCUSSION

Types of Cells Review with students the differences between prokaryotic cells and eukaryotic cells. Prokaryotic cells do not have nuclei, and their genetic material is not separated from the rest of the cell by a membrane. Eukaryotic cells have nuclei, and their genetic material is separated from the rest of the cell by the nuclear membrane. Remind students that prokaryotic cells are usually much smaller than eukaryotic cells.

Answers to MATHBREAK

30 × 4,000 = 120,000 bacteria

The answers to the second question will vary and depend on the size of the classroom.

Terms to Learn

binary fission
endospore

What You'll Do

◆ Describe the characteristics of a prokaryotic cell.
◆ Explain how bacteria reproduce.
◆ Compare and contrast eubacteria and archaebacteria.

Bacteria

Bacteria are the smallest and simplest organisms on the planet. They are also the most abundant. A single gram of soil (which is about equal to the mass of your pencil eraser) can contain over 2.5 billion bacteria!

Not all bacteria are that small. The largest known bacteria are a thousand times larger than the *average* bacterium. Can you imagine an animal a thousand times larger than you? The first giant bacteria ever identified were found in the intestines of a surgeonfish like the one in **Figure 1.**

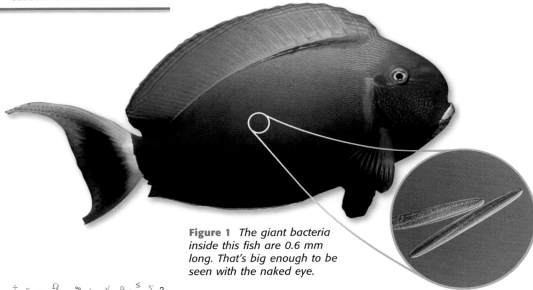

Figure 1 *The giant bacteria inside this fish are 0.6 mm long. That's big enough to be seen with the naked eye.*

MATH BREAK

Airborne Organisms

Air typically has around 4,000 bacteria per cubic meter. Cindy's bedroom is 3 m long and 4 m wide. Her ceiling is 2.5 m high. That means there are 30 m³ in her bedroom (3 m × 4 m × 2.5 m). About how many bacteria are in her bedroom's air? About how many bacteria are in the air of your classroom?

Classifying Bacteria

All organisms fit into one of the six kingdoms: Protista, Plantae, Fungi, Animalia, Eubacteria, and Archaebacteria. Bacteria make up the kingdoms Eubacteria (YOO bak TIR ee uh) and Archaebacteria (AHR kee bak TIR ee uh). These two kingdoms contain the oldest forms of life on Earth. In fact, for over 2 billion years, they were the *only* forms of life on Earth.

Look, Mom! No Nucleus! Bacteria are single-celled organisms that do not have nuclei. A cell with no nucleus is called a *prokaryote*. A prokaryote is able to use cellular respiration, move around, and reproduce. Because a prokaryote has these abilities, it can function as an independent organism.

internet connect

SCI
LINKS
NSTA

TOPIC: Bacteria
GO TO: www.scilinks.org
sciLINKS NUMBER: HSTL230

IS THAT A FACT!

Bacteria are small, but viruses are even smaller. Millions of viruses can fit inside a single bacterium.

Bacterial Reproduction

Most bacteria reproduce by a type of simple cell division known as **binary fission,** illustrated in **Figures 2** and **3.** In binary fission, a prokaryote's DNA is replicated before cell division. The DNA and its copy attach to the inside of the cell membrane. As the cell grows and the membrane grows longer, the loops of DNA become separated. When the cell is about double in size, the membrane pinches inward. A new cell wall forms, separating the two new cells and their DNA.

Figure 2 Binary Fission

❶ The cell grows.

❷ The DNA replicates and attaches to the cell membrane.

❸ The DNA and its copy separate as the cell grows even larger.

❹ The cell splits in two. Each new cell has a copy of the DNA.

Figure 3 *This bacterium is about to complete binary fission.*

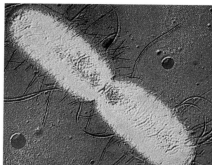

Endospores Each species of bacteria reproduces best at a certain temperature and with a certain amount of moisture. Most species thrive in warm, moist environments. If the environment is unfavorable, some species will be unable to survive. Others will survive by growing a thick protective membrane. These bacteria are then called **endospores.**

Many endospores can survive boiling, freezing, and extremely dry environments. When conditions become favorable again, the endospores will break open, and the bacteria will become active. Scientists have found endospores in the digestive tract of an insect that had been preserved in amber for 30 million years. A similar piece of amber can be seen in **Figure 4.** When the endospores were moistened in a laboratory, the bacteria began to grow!

Figure 4 *Endospores found in a preserved insect indicate that bacteria can survive for millions of years.*

25

CONNECT TO
PHYSICAL SCIENCE

Have students look up the term *fission* in a dictionary. Students should recognize at least two definitions:

• biological definition (cell division)

• physical science definition (the splitting of an atomic nucleus)

2 Teach

READING STRATEGY

Prediction Guide Before students read this page, ask the following question:

What do you think happens to the DNA inside a bacterium before it reproduces? (Accept all reasonable answers. Students should infer that, as with mitosis in eukaryotic cells, the DNA replicates before the cell divides.)

Have students evaluate their answer after they read about bacterial reproduction.

MATH and MORE

Some species of bacteria undergo binary fission every 30 minutes. For such bacteria, have students calculate how many bacteria there would then be after 1 hour, beginning with a single bacterium (4), after 2 hours (16), after 3 hours (64), after 4 hours (256), and after 5 hours (1,024).

USING THE FIGURE

Discuss the stages of bacterial cell division by binary fission, shown in **Figure 2.** Point out to students what takes place during each step of this process. As you review each step, point out the key structures involved.
Sheltered English

Teaching Transparency 37 "Bacterial Reproduction: Binary Fission"

Directed Reading Worksheet Section 1

QuickLab

MATERIALS

FOR EACH GROUP:
• microscope
• prepared slides of bacteria

Safety Caution: Caution students to be careful when handling microscope slides.

Answers to QuickLab

Students should see rod-shaped, spherical, and spiral-shaped bacteria. These are called bacilli, cocci, and spirilla, respectively. The drawings in the ScienceLog should resemble each of these shapes.

DEMONSTRATION

Bacterial Shapes On an overhead projector, display objects that resemble the three shapes of eubacteria, such as a marble, a grain of rice, and a piece of rotini pasta. Ask students to compare the three shapes with the three shapes of eubacteria, illustrated in **Figure 6.** Sheltered English

Homework

Research Types of Bacteria
Have students use the library or the Internet to research the differences in the structure of the cell wall between Gram-positive and Gram-negative bacteria. Students should discover that when bacteria are stained with a specific purple dye and iodine and then rinsed with alcohol and counterstained with safranine, a red dye, some bacteria are stained purple (Gram-positive) and others are stained red (Gram-negative).

Teaching Transparency 38
"The Most Common Shapes of Bacteria"

QuickLab

Spying on Spirilla

Use a **microscope** to observe prepared **slides of bacteria.** What shapes can you see? What are these shapes called? Draw and label the bacteria in your ScienceLog.

The Shape of Bacteria

Almost all bacteria have a rigid cell wall that gives the organism its characteristic shape. Bacteria have a great variety of shapes. The three most common shapes are illustrated below. Each shape provides a different advantage. Can you guess what the advantage of each shape might be?

Whipping Something Up Some bacteria have hairlike structures called flagella (singular, *flagellum*) that help them move around. *Flagellum* means "whip" in Latin. A flagellum spins like a corkscrew, propelling a bacterium through liquid.

The Most Common Shapes of Bacteria

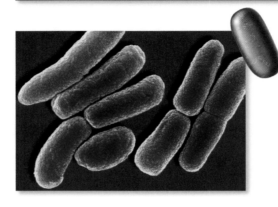

◄ **Bacilli** (buh SIL ie) *are rod-shaped bacteria. Rod-shaped bacteria have a large surface area, which helps them absorb nutrients, but they can also dry out easily.*

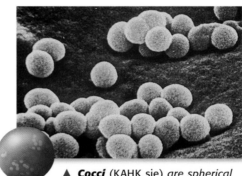

▲ **Cocci** (KAHK sie) *are spherical bacteria. They are more resistant to drying out than rod-shaped bacteria.*

▲ **Spirilla** (spie RI luh) *are long, spiral-shaped bacteria. This is the least common shape for a bacterium. Spirilla move easily in a corkscrew motion, using flagella at both ends.*

WEIRD SCIENCE

Bacteria are used to mine copper. Copper ore contains metal sulfides. The bacteria *Thiobacillus ferrooxidans* take in the sulfides—molecules made of copper and sulfur—and separate the two elements. The bacteria then excrete the purified copper as a waste product.

Kingdom Eubacteria

Most bacteria are eubacteria. The kingdom Eubacteria has more individual organisms than any of the other five kingdoms. Eubacteria have existed for over 3.5 billion years.

Eu Are What Eu Eat Eubacteria are classified by the way they get food. *Consumers* obtain nutrients from other organisms. Most eubacteria, like those helping to decay the leaf in **Figure 5,** are consumers. Many consumers are *decomposers,* which feed on dead organic matter. Other consumers are *parasitic,* which means they invade the body of another organism to obtain food.

Eubacteria that make their own food are *producers*. Some producers are photosynthetic. Like green plants, they convert the energy of the sun into food. These bacteria contain the green pigment *chlorophyll* that is needed for photosynthesis.

Plant Predecessor? Some bacterial producers are *cyanobacteria* (SIE uh noh bak TIR ee uh). Cyanobacteria live in many different types of water environments, such as the one shown in **Figure 6.** It may be that billions of years ago photosynthetic bacteria similar to these began to live inside certain cells with nuclei. The photosynthetic bacteria made food, and the host provided a protected environment for the bacteria. This might be how the first plants came to be.

Figure 5 *Decomposers, like the ones helping to decay this leaf, return nutrients to the ecosystem.*

✓ Self-Check

Cyanobacteria were once classified as plants. Can you explain why? *(See page 168 to check your answer.)*

Figure 6 *Cyanobacteria in this hot spring excrete chemicals that help form the chalky terraces.*

27

③ Extend

RESEARCH

Bacterial Environments Ask students to use the library or the Internet to research the extreme habitats of archaebacteria. Encourage students to make posters of their findings to share with the class.

Homework

Concept Mapping After students read this section, have them create a concept map in their ScienceLog that includes the following terms:

bacteria, Eubacteria, Archaebacteria, prokaryotes, nucleus, cells, eukaryotes, harsh environments

Answers to Self-Check

Cyanobacteria were once classified as plants because they use photosynthesis to make food.

BRAIN FOOD

Thiomargarita namibiensis is a species of bacteria so large, an individual bacterium is visible to the naked eye. It is the largest known bacterium, about 750 microns (0.030 in.) wide. Even large bacteria are single-celled organisms, and their surface-area-to-volume ratio is far less than that of smaller bacteria. What problems might this cause an organism? (A small surface-area-to-volume ratio makes it difficult to take in a sufficient quantity of nutrients and eliminate wastes efficiently.)

Quiz

1. **What functions must a prokaryotic cell perform?**
(Prokaryotic cells must respire, make or consume food, move around on their own, and reproduce.)

2. **What are the three main types of archaebacteria?**
(methane makers, heat lovers, and salt lovers)

ALTERNATIVE ASSESSMENT

 Divide the class into two teams, and provide each team with 10 large index cards. Have each team look through this section and write 10 questions about the characteristics of bacteria, one question per index card. Bring the teams together to play a trivia game about bacteria.

Reinforcement Worksheet
"Bacteria Bonanza"

---

internetconnect

SCiLINKS
NSTA

TOPIC: Archaebacteria
GO TO: www.scilinks.org
*sci*LINKS NUMBER: HSTL235

Kingdom Archaebacteria

Archaebacteria thrive in places where no other living things are found. Scientists have found archaebacteria in the hot springs at Yellowstone National Park and beneath 430 m of ice in Antarctica. They have even been found living 8 km below the Earth's surface!

Archaebacteria are genetically different from eubacteria. Not all archaebacteria have cell walls. The cell walls of archaebacteria—when they do have them—are chemically different from those of all other organisms.

Pass the Salt There are three main types of archaebacteria: methane makers, heat lovers, and salt lovers. Methane makers excrete methane gas. They are found in many places including swamps. Heat lovers live in places like ocean rift vents where temperatures are over 360°C. Salt lovers live in places where the concentration of salt is very high, such as the Dead Sea, shown in **Figure 7**.

Figure 7 *The Dead Sea is so salty that only archaebacteria can survive in it. Fish carried into the Dead Sea by the Jordan River die instantly.*

internetconnect

SCiLINKS
NSTA

TOPIC: Bacteria, Archaebacteria
GO TO: www.scilinks.org
*sci*LINKS NUMBER: HSTL230, HSTL235

SECTION REVIEW

1. Draw and label the three main shapes of bacteria.

2. Describe the four steps of binary fission.

3. How do eubacteria and archaebacteria differ?

4. **Analyzing Concepts** Many bacteria cannot reproduce in cooler temperatures and are destroyed at high temperatures. How do humans take advantage of this when preparing and storing food?

28

▼ **Answers to Section Review**

1. The drawings should resemble the pictures on page 26.

2. The cell grows, the DNA replicates, the DNA molecules separate as the cell grows, and the cell splits in two.

3. Eubacteria and archaebacteria are genetically different. Their cell walls are also chemically different.

4. Humans store food in refrigerators and freezers to slow bacterial growth. Humans also cook food at high temperatures, which helps kill many bacteria.

Terms to Learn

bioremediation
antibiotic
pathogenic bacteria

What You'll Do

◆ Explain why life on Earth depends on bacteria.
◆ List five ways bacteria are useful to people.
◆ Describe why some bacteria are harmful to people.

Bacteria's Role in the World

Bacteria may be invisible to us, but their effects on the planet are not. Because many types of bacteria cause disease, bacteria have gotten a bad reputation. However, they also do many things that are important to humans.

Good for the Environment

Life as we know it could not exist without bacteria. They are vital to our environment, and we benefit from them in several ways.

Nitrogen-Fixing   Nitrogen is an essential chemical for all organisms because it is a component of proteins and DNA. Plants must have nitrogen in order to grow properly. You might think getting nitrogen would be easy because nitrogen gas makes up more than 75 percent of the air. But most plants cannot use nitrogen from the air. They need to take in a different form of nitrogen. *Nitrogen-fixing bacteria* consume nitrogen in the air and change it into a form that plants can use. This process is described in **Figure 8.**

Figure 8   Bacteria's Role in the Nitrogen Cycle.

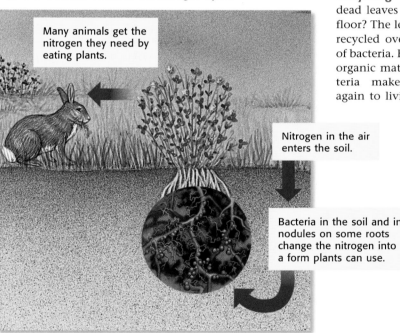

Many animals get the nitrogen they need by eating plants.

Nitrogen in the air enters the soil.

Bacteria in the soil and in nodules on some roots change the nitrogen into a form plants can use.

Recycling   Have you ever seen dead leaves and twigs on a forest floor? The leaves and twigs will be recycled over time with the help of bacteria. By breaking down dead organic matter, decomposing bacteria make nutrients available again to living things.

29

IS THAT A FACT!

Food crops use nitrogen very rapidly. To make sure that crops receive all the nitrogen they need, farmers worldwide applied more than 65 million metric tons of nitrogen fertilizer in 1990.

 **LabBook**   **PG 136**

Aunt Flossie and the Intruder

Focus

Bacteria's Role in the World

This section explains why life on Earth depends on bacteria. Students learn how bacteria are both harmful and beneficial to people.

🔔 Bellringer

Display the following question on the board or an overhead projector:

Are harmful bacteria more of a problem or less of a problem to people now than they were 200 years ago? (Students should infer that harmful bacteria generally cause fewer problems because people understand that maintaining hygienic conditions eliminates many bacteria. The discovery of antibiotics has also helped us overcome some of the health problems posed by bacteria.)

1 Motivate

DISCUSSION

Bacterial Products   Ask students to name products they use that are made using bacteria. (Answers might include yogurt, cheese, sauerkraut, and other foods.)

Ask students what products exist because of the scientific discovery that bacteria can cause disease. (Answers might include sterile bandages, antiseptic creams and soaps, antibacterial cleansers, canned food, antibiotics, and water filters.)

 Directed Reading Worksheet Section 2

SCIENTISTS AT ODDS

Joseph Lister (1827–1912), a professor of surgery at Glasgow University, in Scotland, pioneered using antiseptics in surgery. Surgeons of the day were critical of Lister. But the surgeons who followed Lister's techniques immediately had greater success in surgery; the death rate fell from 1 in 3 to 1 in 20. In time, all surgeons were educated about the importance of using antiseptics in surgery and hospital care. Operations today are much safer because of one man's determination to save his patients.

REAL-WORLD CONNECTION

Fresh foods are susceptible to bacteria, but canning food enables us to keep food year-round. It also makes food more convenient. In 1795, Napoleon offered a prize for the most practical method of preserving food for soldiers. In 1804, a chef named Nicolas Appert discovered that liquids, such as soups, and small fruits, like cherries, could be preserved by packing them into champagne bottles and plunging the sealed bottles into a bath of boiling water. Although he didn't know why at the time, he was sterilizing bacteria that would have otherwise spoiled the food.

Interactive Explorations CD-ROM "Scope It Out!"

internetconnect

SCLINKS
NSTA

TOPIC: Antibiotics
GO TO: www.scilinks.org
sciLINKS NUMBER: HSTL240

Figure 9 *Bioremediating bacteria are added to soil to consume pollutants and excrete them as harmless chemicals.*

BRAIN FOOD

Alexander Fleming, the Scottish scientist who discovered antibiotics, created a microbial growth shaped like the British flag in honor of the queen's visit to his lab. She was *not* amused.

Cleaning Up Recently bacteria have been used to combat pollution. **Bioremediation** (BIE oh ri MEE dee AY shuhn) is the use of bacteria and other microorganisms to change pollutants into harmless chemicals. Bioremediation is used to clean up industrial, agricultural, and municipal wastes, as well as oil spills. The workers in **Figure 9** are using bioremediation to remove toxins from the soil.

Figure 10 *Genes from the* Xenopus *frog were used to produce the first genetically engineered bacteria.*

Good for People

Scientists are constantly searching for new ways to use bacteria to better the lives of humans. People have been able to genetically engineer bacteria since 1973. It was then that researchers inserted genes from a frog like the one in **Figure 10** into the bacterium *Escherichia coli* (ES uhr RI shee uh COHL ie). The bacterium started reproducing the frog genes. Never before had such a genetically altered organism existed.

Now scientists can genetically engineer bacteria for many purposes, including the production of medicines, insecticides, cleansers, adhesives, foods, and many other products.

Fighting Bacteria with Bacteria Although some bacteria cause diseases, others make chemicals that treat diseases. **Antibiotics** are medicines used to kill bacteria and other microorganisms. Many bacteria have been genetically engineered to make antibiotics in large quantities.

30

Multicultural CONNECTION

One of the first people to describe the importance of composting dead plant material was African-American agricultural chemist George Washington Carver (1864–1903). While teaching at Tuskegee Institute, Carver and his students added leaves, weeds, and potato peelings to the soil. Once these materials were incorporated, bacteria in the soil broke down the plant material into nutrients that plants could use.

Ingenious Engineering!

Ralph specializes in genetically engineering bacteria. He has engineered many different types of bacteria to help solve medical and environmental problems, but today he is completely out of ideas. He needs you to help him think of a design for a new bacterium that will either treat a disease or help the environment. What would you want this new bacterium to do? What kinds of traits would you give it?

Insulin Scientists have created genetically engineered bacteria that can produce other medicines, such as insulin. Insulin is a substance needed by the body to properly use sugars and other carbohydrates. People who have *diabetes* cannot produce the insulin they need. They must take insulin daily. In the late 1970s, scientists put genes carrying the genetic code for human insulin into *E. coli* bacteria. The bacteria produced human insulin, which can be separated from the bacteria and given to diabetics.

Feeding Time! Believe it or not, people breed bacteria for food! Every time you eat cheese, yogurt, buttermilk, or sour cream, you also eat a lot of lactic-acid bacteria. *Lactic-acid bacteria* digest the milk sugar lactose and convert it into lactic acid. The lactic acid acts as a preservative and adds flavor to the food. The foods in **Figure 11** could not be made without bacteria.

Activity

Create a week's meal plan without any foods made with bacteria. What would your diet be like without prokaryotes?

TRY at HOME

Figure 11 *Bacteria are used to make many different types of food, including sauerkraut, sourdough bread, some kinds of sausages, pickles, and dairy products.*

31

IS THAT A FACT!

A bulging can of food is a dangerous object! The bulge could be indicative of the presence of the *Clostridium botulinum* bacteria, which can cause a severe type of food poisoning called botulism. Caution students never to eat food from a can with a bulging lid or sides.

Answers to APPLY

All logical answers should be accepted.

ACTIVITY

The Spread of Bacteria Before students come to class, spread the dust from colored chalk on objects the students will touch. Also cover your hands with chalk dust. Without telling students why, encourage them to participate in activities that will spread the dust around the room and from person to person. Ask students to describe how this activity models the way bacteria are spread and to offer solutions to stop the distribution of bacteria. Sheltered English

DISCUSSION

Bacterial Illnesses Ask students to raise their hand if they have ever had strep throat, food poisoning, or bacterial pneumonia. Have students describe how they felt while they were sick and to describe any medical treatment they received.

Answer to Activity

Answers will vary.

Homework

Writing **Research/Oral Report** Have students research the role of bacteria in the cleanup efforts after the massive oil spill from the oil tanker *Exxon Valdez*. The spill occurred in 1989 in Prince William Sound, in Alaska. Have students share their findings in oral reports they present to the class.

PORTFOLIO

Quiz

1. Explain the role of bacteria in recycling. (Bacteria act as decomposers of dead organisms. They break down the materials in the dead organisms so that nutrients will be available for use by other living things.)

2. Explain why bacteria are important in helping plants obtain nitrogen. (Nitrogen-fixing bacteria that live in nodules on a plant's roots consume nitrogen gas and change it into a form that can be used by plants.)

ALTERNATIVE ASSESSMENT

Writing On the board, write the headings "Helpful Bacteria" and "Harmful Bacteria." Have student volunteers go to the board one at a time and write an example of how bacteria can be either helpful or harmful. Review the list as a class.

MISCONCEPTION ///ALERT

The man in the bed in **Figure 12** is depicted in blue. This is to symbolize that he had the bubonic plague. People with the plague do not actually appear blue.

Critical Thinking Worksheet "Bacterial Blastoff"

Harmful Bacteria

We couldn't survive without bacteria, but they are also capable of doing incredible damage. Scientists realized in the mid-1800s that some bacteria are pathogenic. **Pathogenic bacteria** cause diseases, such as the one illustrated in **Figure 12.** These bacteria invade a host organism and obtain nutrients from the host's cells. In the process, they cause damage to the host. Today, almost all bacterial diseases can be treated with antibiotics. Many can also be prevented with vaccines. Some diseases caused by bacteria are shown in the table below.

Figure 12 *Between the years 1346 and 1350, the bubonic plague killed 25 million people. That was one-third of Europe's population at the time.*

Bacterial Diseases

- Dental cavities
- Ulcers
- Strep throat
- Food poisoning
- Bacterial pneumonia
- Lyme disease
- Tuberculosis
- Leprosy
- Typhoid fever
- Bubonic plague

Enough Diseases to Go Around Bacteria cause diseases in other organisms as well as in people. Have you ever seen a plant with discolored spots or soft rot? If so, you've seen bacterial damage to another organism.

Pathogenic bacteria attack plants, animals, protists, fungi, and even other bacteria. They can cause considerable damage to grain, fruit, and vegetable crops. The branch of a pear tree in **Figure 13** shows the effects of pathogenic bacteria.

Figure 13 *This branch of a pear tree has fire blight, a bacterial disease.*

SECTION REVIEW

1. List three different products bacteria are used to make.
2. What are two ways that bacteria affect plants?
3. How can bacteria both cause and cure diseases?
4. **Analyzing Relationships** Describe some of the problems humans would face if there were no bacteria.

▼ **Answers to Section Review**

1. Sample answer: insulin, pickles, adhesives

2. Bacteria affect plants by providing them with nitrogen (nitrogen-fixing bacteria) and by providing them with nutrients from decayed organic matter (decomposers).

3. Pathogenic bacteria cause disease by obtaining nutrition from their host's cells. Other bacteria can be used to make antibiotics that help kill pathogenic bacteria. Many types of bacteria also make medicines, such as insulin.

4. Answers can vary. Sample answer: If there were no bacteria, we would not have pickles, cheese, or yogurt. Also, nothing would decompose, and plants wouldn't be able to grow because they wouldn't be able to process nitrogen.

Viruses

What You'll Do

◆ Explain how viruses are similar to and different from living things.
◆ List the four major virus shapes.
◆ Describe the two kinds of viral reproduction.

Viruses have been called the greatest threat to the survival of humanity. But what are they? A **virus** is a microscopic particle that invades a cell and often destroys it. They are everywhere, and for humans they are mostly one big headache. That's because many diseases are caused by viruses, including the common cold, flu, and acquired immune deficiency syndrome (AIDS). AIDS is caused by the human immunodeficiency virus (HIV).

It's a Small World

Viruses are incredibly tiny. They are even smaller than the smallest bacteria. About 5 billion of them can fit into a single drop of blood. Because of viruses' small size and ever-changing nature, scientists don't know how many types of viruses exist. The number may be in the billions or higher!

Are They Living?

Like living things, viruses contain protein and nucleic acids. But viruses, such as the ones shown in **Figure 14,** don't eat, grow, breathe, or perform other biological functions. A virus cannot "live" on its own, although it can reproduce inside a living organism that serves as its host. A **host** is an organism that supports a parasite. Using a host's cell as a miniature factory, viruses instruct the cell to produce viruses rather than healthy new cells.

Focus

Viruses

This section describes the characteristics that viruses and living things share, and those they do not. Students will learn that viruses are classified by the disease they cause, what they do inside host cells, or by their shape. Students will learn how viruses reproduce.

Bellringer

Asks students to answer the following question in their ScienceLog:

What are the characteristics of living things? (In Chapter 1, students learned that all living things have cells, sense and respond to change, reproduce, have DNA, use energy, and grow and develop.)

MATH BREAK

Sizing Up a Virus

If you enlarged an average virus 600,000 times, it would be about the size of a small pea. How tall would you be if you were enlarged 600,000 times?

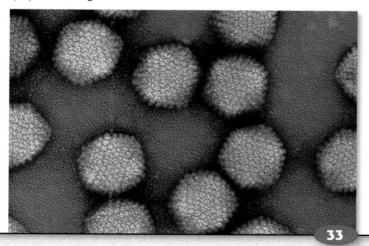

Figure 14 *Viruses are not cells. They do not have cytoplasm or organelles.*

Answer to MATHBREAK

This answer will vary depending on the student's height. Example: If the student is 1.6 m (5 ft 2 in.), he or she would be 960,000 m (3,100,000 ft) tall if enlarged 600,000 times.

1) Motivate

DEMONSTRATION

Flame Tulips Show students a flame tulip or pictures of a flame tulip. Its name is due to the distinctive contrasting colors in its petals. (Photos of these beautiful tulips can be found in bulb and seed catalogs.) Explain to students that a virus that infects the flower causes the distinctive colors. These mutant plants were hybridized by plant breeders and are grown in gardens around the world. Sheltered English

Directed Reading Worksheet Section 3

RESEARCH

Viral Diseases After students read this section, have them conduct library or Internet research to find out about a virus that causes the disease, the symptoms of the disease, and the treatment for the disease. Have students prepare an oral report about their findings to share with the class.

CONNECT TO
EARTH SCIENCE

Crystallography is a branch of chemistry in which scientists study the specific form and structure of crystals. Like specific viruses that always form the same type of crystals, other materials also form crystals that have a distinct size, shape, and number of facets. For example, NaCl, table salt, forms crystals visible with a magnifying glass. Use the following Teaching Transparency to illustrate the crystalline structure of gold.

Teaching Transparency 107
"Gold Crystal Structure"
LINK TO EARTH SCIENCE

Teaching Transparency 39
"The Basic Shapes of Viruses"

📶 internet**connect**

*SCi*LINKS
NSTA

TOPIC: Viruses
GO TO: www.scilinks.org
*sci***LINKS NUMBER:** HSTL245

Chemistry
C O N N E C T I O N

Many viruses can form crystalline structures. This is a property of chemicals, not cellular organisms.

Classifying Viruses

Viruses can be grouped by the type of disease they cause, their life cycle, or the type of genetic material they contain. Viruses can also be classified by their basic shape, as illustrated below. No matter what its structure is, every virus is basically some form of genetic material enclosed in a protein coat.

The Basic Shapes of Viruses

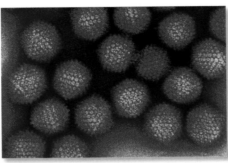

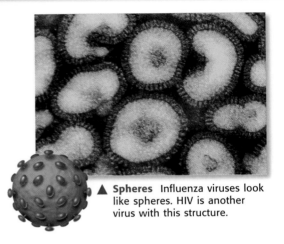

▲ **Crystals** The polio virus is shaped like the crystals shown here.

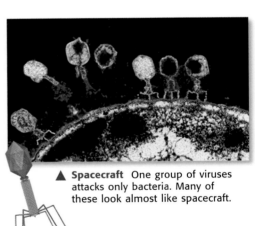
▲ **Spheres** Influenza viruses look like spheres. HIV is another virus with this structure.

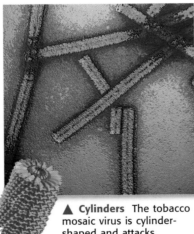

▲ **Cylinders** The tobacco mosaic virus is cylinder-shaped and attacks tobacco plants.

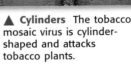
▲ **Spacecraft** One group of viruses attacks only bacteria. Many of these look almost like spacecraft.

A Destructive House Guest

The one function that viruses share with living things is that they reproduce. They do this by infecting living cells and turning them into virus factories. This is called the *lytic cycle,* as shown in **Figure 15.**

Figure 15 The Lytic Cycle

1 The virus finds a host cell.

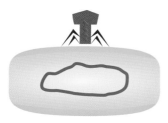

2 The virus enters the cell, or in some cases, the virus's genes are injected into the cell.

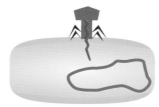

3 Once the virus's genes are inside, they take over the direction of the host cell, turning it into a virus factory.

4 The new viruses break out of the host cell ready to find a new host and repeat the cycle.

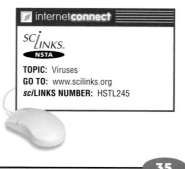

A Ticking Time Bomb   Some viruses don't go straight into the lytic cycle. These viruses insert their genes into the host cell, but no new viruses are made immediately. When the host cell divides, each new cell has a copy of the virus's genes. This is called the *lysogenic cycle.* The viral genes can remain inactive for long periods of time until a change in the environment or stress to the organism causes the genes to launch into the lytic cycle.

SECTION REVIEW

1. What would happen if one generation of measles viruses never found a host?

2. Describe the four steps in the lytic cycle.

3. **Analyzing Relationships**   Do you think modern transportation has had an effect on the way viruses are spread? Explain your answer.

internet**connect**

SC**LINKS**
NSTA

TOPIC: Viruses
GO TO: www.scilinks.org
*sci***LINKS NUMBER:** HSTL245

Quiz

1. Describe three shapes of viruses, and give an example of each. (Sample answer: crystals—polio virus; spheres—influenza virus; rods—tobacco mosaic virus)

2. How are viruses like and unlike living things? (Like living things, viruses contain protein and nucleic acids. But unlike living things, viruses don't eat, grow, breathe, or perform other functions.)

ALTERNATIVE ASSESSMENT

Writing Have students write a short story in which they discover a new type of disease-causing virus. Ask students to include specific details about the structure of the new virus and how it reproduces. Ask student volunteers to read their stories to the class.

Teaching Transparency 40
"The Lytic Cycle"

Reinforcement Worksheet
"The Lytic Cycle"

▼ Answers to Section Review

1. If a generation of measles viruses never found a host, measles would die out; viruses can't reproduce themselves without a host.

2. First, the virus enters the host cell. Second, the virus releases its genes into the host cell. Third, the host cell is directed to make more viral genes, and new proteins are made to make coats for the new viruses. Fourth, the host cell is destroyed and the viruses move to infect other cells.

3. Answers can vary but should be logical. People and livestock are much more mobile now, and thus able to bring viruses to new places.

Viral Decorations
Teacher's Notes

Time Required
One 45-minute class period

Lab Ratings

EASY ——————→ HARD

TEACHER PREP   🔺🔺
STUDENT SET-UP   🔺🔺🔺
CONCEPT LEVEL   🔺🔺🔺🔺
CLEAN UP   🔺🔺

Safety Caution
Remind students to review all safety cautions and icons before beginning this lab activity.

Gladys Cherniak
St. Paul's Episcopal School
Mobile, Alabama

Making Models Lab

Viral Decorations

It's true that viruses are made of only protein and genetic material. But their structures have many different shapes. The shapes help the viruses attach to and get inside of living cells. One viral shape can be made from the template on page 37. In this activity, you will make and modify a model of a virus.

MATERIALS
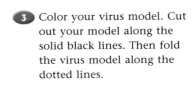

- virus model template
- construction paper
- colored markers
- scissors
- glue or tape
- pipe cleaners, twist ties, buttons, string, plastic wrap, and recycled or other reusable materials for making variations of the virus

Procedure

1. Carefully copy the virus model template on the next page onto a piece of construction paper. You may make the virus model as large as your teacher allows.

2. Do some research on viruses that have a shape similar to those on this page. Decide how you will change your model. For example, you might want to add the tail and tail fibers of a spacecraft-shaped virus. Or you might wrap the model in plastic to represent the envelope that surrounds the protein coat in HIV.

3. Color your virus model. Cut out your model along the solid black lines. Then fold the virus model along the dotted lines.

Bacteriophage

Influenza
Human Immunodeficiency Virus (HIV)

36

---

4. Glue or tape each lettered tab under the corresponding lettered triangle. For example, glue or tape the large Z tab under the Z-shaded triangle. When you are done, you should have a closed box with 20 sides.

5. Make the changes that you planned in step 2. Write the name of your virus on the model. Decorate your classroom with your virus and those of your classmates.

Analysis

6. Describe the changes you made to your virus model. How do you think the virus might use them?

7. Does your virus cause disease? If so, explain what disease it causes, how it reproduces, and how the virus is spread.

Going Further

Research in the library or on the Internet an unusual virus that causes an illness, such as the influenza virus, HIV, or Ebola virus. Explain what is unusual about the virus, what illness it causes, and how it might be avoided.

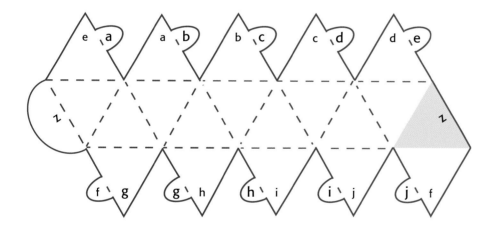

Answers
6. Answers will vary.
7. Answers will vary.

Going Further
The influenza virus, of course, causes flu. HIV is the virus that causes AIDS, and Ebola-Zaire is a rare filovirus that causes a deadly hemorrhagic fever.

 Datasheets for LabBook

Chapter Highlights

VOCABULARY DEFINITIONS

SECTION 1

endospore a membrane containing inactive cell material

binary fission the simple cell division in which one cell splits into two; used by bacteria

Chapter Highlights

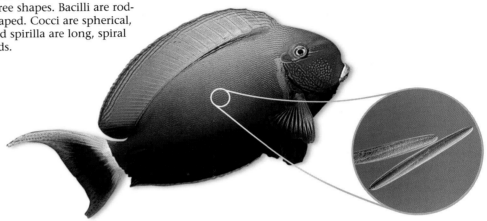

SECTION 1

Vocabulary
    binary fission *(p. 25)*
    endospore *(p. 25)*

Section Notes

- Bacteria are classified as eubacteria or archaebacteria. Both are prokaryotes.

- Most bacteria have one of three shapes. Bacilli are rod-shaped. Cocci are spherical, and spirilla are long, spiral rods.

- Most bacteria reproduce by binary fission. Some produce endospores that can survive for thousands of years.

- Many eubacteria are decomposers. They obtain their energy from dead organic matter. Others are parasites.

- Many bacterial producers, including cyanobacteria, contain chlorophyll.

- Archaebacteria live in places where no other organisms can survive. Many are grouped as salt lovers, methane makers, and heat lovers.

☑ Skills Check

Math Concepts

MULTIPLYING MICROORGANISMS Some bacteria can divide every 20 minutes in an ideal growing environment. That means that if you start out with just one bacterium, in an hour there will be eight bacteria. There will be 512 bacteria in 3 hours, 32,768 in 5 hours, and 1,073,741,824 bacteria in 10 hours!

Visual Understanding

BACTERIA VERSUS VIRUSES The diagrams on pp. 25 and 35 illustrate the way some bacteria and viruses reproduce. Make sure you understand how each process works. Viruses use their hosts' cells to reproduce. Think about how this is different from the way pathogenic bacteria use a host's cell.

38

Lab and Activity Highlights

Viral Decorations `PG 36`

Aunt Flossie and the Intruder `PG 136`

 Datasheets for LabBook (blackline masters for these labs)

SECTION 2

Vocabulary

bioremediation *(p. 30)*
antibiotic *(p. 30)*
pathogenic bacteria *(p. 32)*

Section Notes

- Bacteria are important to the planet. Some act as decomposers. Others convert nitrogen gas to a form that plants can use.

- Bacteria are used in making a variety of foods, medicines, and pesticides. They are also used to clean up pollution.

- Pathogenic bacteria cause diseases in humans as well as other organisms.

Labs

Aunt Flossie and the Intruder *(p. 136)*

SECTION 3

Vocabulary

virus *(p. 33)*
host *(p. 33)*

Section Notes

- Viruses have characteristics of both living and nonliving things. They can reproduce only inside a living cell.

- Viruses may be classified by their structure, the kind of disease they cause, or their life cycle.

- In order for a virus to reproduce, it must enter a cell, reproduce itself, and then break open the cell. This is called the lytic cycle.

- The genes of a virus are incorporated into the genes of the host cell in the lysogenic cycle. The virus's genes may remain inactive for years.

VOCABULARY DEFINITIONS, *continued*

SECTION 2

bioremediation the use of bacteria and other microorganisms to change pollutants in soil and water into harmless chemicals

antibiotic substance used to kill bacteria or slow the growth of bacteria

pathogenic bacteria bacteria that invade a host organism and obtain the food they need for growth and reproduction from the host's cells

SECTION 3

virus a microscopic particle that invades a cell and often destroys it

host an organism on which a parasite lives

Vocabulary Review Worksheet

Blackline masters of these Chapter Highlights can be found in the **Study Guide.**

internet connect

GO TO: go.hrw.com

Visit the **HRW** Web site for a variety of learning tools related to this chapter. Just type in the keyword:

KEYWORD: HSTVIR

SCI LINKS

N S T A

GO TO: www.scilinks.org

Visit the **National Science Teachers Association** on-line Web site for Internet resources related to this chapter. Just type in the *sci*LINKS number for more information about the topic:

TOPIC: Bacteria	*sci*LINKS NUMBER: HSTL230
TOPIC: Archaebacteria	*sci*LINKS NUMBER: HSTL235
TOPIC: Antibiotics	*sci*LINKS NUMBER: HSTL240
TOPIC: Viruses	*sci*LINKS NUMBER: HSTL245

39

Lab and Activity Highlights

LabBank

Labs You Can Eat, Bacterial Buddies

Inquiry Labs, It's an Invasion!

EcoLabs & Field Activities, Ditch's Brew

Long-Term Projects & Research Ideas, Bacteria to the Rescue!

Interactive Explorations CD-ROM

CD 1, Exploration 3, "Scope It Out!"

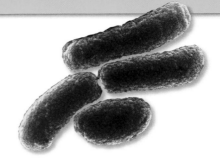

Chapter Review
Answers

USING VOCABULARY

1. bacilli
2. binary fission
3. antibiotics
4. host
5. nitrogen-fixing
6. Producers

UNDERSTANDING CONCEPTS

Multiple Choice

7. d
8. a
9. b
10. b
11. c
12. d
13. c
14. d

Short Answer

15. Both forms of bacteria help plants grow. Nitrogen-fixing bacteria supply plants with nitrogen, and decomposers supply plants with nutrients.
16. In the lytic cycle, the virus immediately reproduces after infecting the host. In the lysogenic cycle, the virus inserts its genes into the host cell. The host cell divides several times before viruses begin their lytic cycle.

Chapter Review

USING VOCABULARY

To complete the following sentences, choose the correct term from each pair of terms listed below:

1. Rod-shaped bacteria are called __?__. (*bacilli* or *cocci*)

2. Most bacteria reproduce by __?__. (*endospores* or *binary fission*)

3. Bacterial infections can be treated with __?__. (*antibiotics* or *bioremediation*)

4. A virus needs a __?__ to reproduce. (*crystal* or *host*)

5. Without __?__ bacteria, life on Earth could not exist. (*pathogenic* or *nitrogen-fixing*)

6. __?__ make their own food. (*Consumers* or *Producers*)

UNDERSTANDING CONCEPTS

Multiple Choice

7. Bacteria are used for all of the following except
   a. making certain foods.
   b. making antibiotics.
   c. cleaning up oil spills.
   d. preserving fruit.

8. In the lytic cycle
   a. the host cell is destroyed.
   b. the host cell destroys the virus.
   c. the host cell becomes a virus.
   d. the host cell undergoes cell division.

9. A bacterial cell
   a. is an endospore.
   b. has a loop of DNA.
   c. has a distinct nucleus.
   d. is a eukaryote.

10. Eubacteria
    a. include methane makers.
    b. include decomposers.
    c. all have chlorophyll.
    d. are all rod-shaped.

11. Cyanobacteria
    a. are consumers.
    b. are parasites.
    c. contain chlorophyll.
    d. are decomposers.

12. Archaebacteria
    a. are a special type of eubacteria.
    b. live only in places without oxygen.
    c. are primarily lactic-acid bacteria.
    d. can live in hostile environments.

13. Viruses
    a. are about the same size as bacteria.
    b. have nuclei.
    c. can reproduce only within a host cell.
    d. don't infect plants.

14. Bacteria are important to the planet as
    a. decomposers of dead organic matter.
    b. processors of nitrogen.
    c. makers of medicine.
    d. All of the above

Short Answer

15. How are the functions of nitrogen-fixing bacteria and decomposers similar?

16. What is the difference between the lytic cycle and the lysogenic cycle?

40

Concept Mapping

17. Use the following terms to create a concept map: eubacteria, bacilli, cocci, spirilla, parasites, consumers, producers, cyanobacteria.

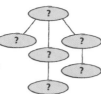

CRITICAL THINKING AND PROBLEM SOLVING

Write one or two sentences to answer the following questions:

18. Describe some of the problems you think bacteria might face if there were no humans.

19. A nuclear power plant explodes and wipes out every living thing within a 30 km radius. What kind of organism do you think might colonize the radioactive area first? Why?

MATH IN SCIENCE

20. An ounce is equal to about 28 g. If 1 g of soil contains 2.5 billion bacteria, how many bacteria are in 1 oz?

21. A bacterial cell infected by a virus divides every 20 minutes. After 10,000 divisions, the virus breaks loose from its host cell. About how many weeks will this take?

INTERPRETING GRAPHICS

The following diagram illustrates the stages of binary fission. Match each statement with the correct stage.

1

2

3

4

22. The DNA loops separate.

23. The DNA loop replicates.

24. The parent cell starts to expand.

25. The DNA attaches to the cell membrane.

Reading Check-up

Take a minute to review your answers to the Pre-Reading Questions found at the bottom of page 22. Have your answers changed? If necessary, revise your answers based on what you have learned since you began this chapter.

Concept Mapping

17. An answer to this exercise can be found at the front of this book.

CRITICAL THINKING AND PROBLEM SOLVING

18. Sample answer: Most bacteria would face no problems at all. The only bacteria that might have a problem are pathenogenic bacteria that only infect humans.

19. Sample answer: Perhaps archae-bacteria would colonize first because they can live in hostile environments.

MATH IN SCIENCE

20. about 70 billion bacteria
21. about 20 weeks:
$$200{,}000 \text{ min} \times \frac{1 \text{ hr}}{60 \text{ min}} \times \frac{1 \text{ day}}{24 \text{ hr}} \times \frac{1 \text{ week}}{7 \text{days}} = 19.8 \text{ or } 20 \text{ weeks}$$

INTERPRETING GRAPHICS

22. stage 3
23. stage 2
24. stage 1
25. stage 2

Concept Mapping Transparency 10

Blackline masters of this Chapter Review can be found in the **Study Guide**.

Background

During the 1970s, molecular biologists developed techniques to manipulate the genetic information of plants and animals. Scientists use enzymes called restriction enzymes to "cut out" and isolate certain genes. These genes are then incorporated into the DNA of other organisms. Human growth hormone and insulin, formerly collected from human cadavers, were among the first medicinal products to be produced using these techniques.

Genes contain sets of instructions that tell the body how to make specific proteins. In the case of transgenic organisms, the new genes instruct the molecular machinery of the organisms to produce new proteins. Encourage students to consider any ethical issues that may surround the production of transgenic animals. Is using animals to produce drugs and organs different from using them to produce food? Are there genes we should not transfer into animals? How many genes does it take before an animal might be considered human?

Science, Technology, and Society
Edible Vaccines

No one likes getting a shot, right? Unfortunately, though, shots are a necessary part of life. This is because people in this country are protected from life-threatening diseases by vaccinations. But vaccination shots are expensive, may require refrigeration, and require trained medical professionals to administer them. These facts often keep people in developing countries from getting vaccinated.

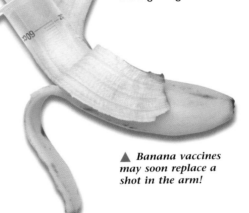

▲ *Banana vaccines may soon replace a shot in the arm!*

Pass the Banana, Please

Edible vaccines would have several important advantages over traditional injected vaccines. First of all, they wouldn't require a painful shot! But more important, these vaccines will be cheaper to produce and may not require refrigeration or trained medical professionals. Plants like bananas are easily grown in fields and greenhouses in many Third World countries. One banana could even carry vaccines for several diseases at one time! But just how could a banana do this?

Add Some DNA

Scientists have made DNA that closely resembles the "fingerprints" of specific disease particles. They can insert this DNA into banana genes that code for proteins. Scientists hope they can trick the human immune system into recognizing these proteins as invaders and producing the necessary antibodies to fight diseases. Unlike traditional vaccines, these transgenic bananas (bananas containing foreign DNA) do not carry the risk of infection because they do not contain any viral particles.

Transgenic plants aren't all that new. Agricultural scientists make transgenic plants to improve crops. Scientists have introduced new DNA into fruits and vegetables so that they are more resistant to pests and drought; have larger, sweeter, and more colorful fruit; and ripen more quickly.

Important Questions

Unfortunately, edible vaccines will not be available for several years. Some safety concerns must still be addressed before these vaccines can be given to people. Do edible vaccines have side effects? How long will the resistance last? What happens if someone eats too many bananas? Research labs are testing the vaccines to answer these questions.

Check It Out!

▶ The milk of transgenic animals is also being tested as a potential vaccine carrier. Scientists hope that goat's milk containing malarialike proteins will prevent as many as 3 million deaths each year. Investigate for yourself how the malaria vaccine will work.

42

Answer to Check It Out!

Scientists put the genes for the proteins in the goat's genome. When the mammary cells express the proteins into the milk, the scientists can then milk the goats and isolate the malarialike proteins. The scientists then can use the proteins to make a vaccine.

Health
Helpful Viruses

L ess than 100 years ago, people had no way to treat bacterial infections. If you became ill from contact with pathogenic bacteria, you could only hope that your immune system would be able to defeat the invaders. But in 1928 a Scottish scientist named Alexander Fleming discovered the first antibiotic, or bacteria-killing drug. This first antibiotic was called penicillin.

Using Viruses to Fight Bacteria
Since Fleming's discovery, people have used antibiotics to treat infections and to purify water supplies. But scientists are now realizing that many bacteria are becoming resistant to existing antibiotics. It is quite possible that the overuse of antibiotics will make all current antibiotics ineffective in the near future. So what will people use to fight bacterial infections? Some scientists think viruses might be the answer! You might be thinking that viruses can only cause diseases, not cure them, but there is a particular type of virus, called a bacteriophage, that attacks only bacteria.

How Do They Do This?
Bacteriophages destroy bacteria cells in the same way other viruses can destroy animal or plant cells. Each kind of bacteriophage can only infect a particular species of bacteria. This can make an extremely effective antibiotic. Existing antibiotics kill not only harmful bacteria but also bacteria that people need to stay healthy. This can make people treated with antibiotics very sick, causing a breakdown in their immune system or digestive process. Because bacteriophages would kill only specific harmful bacteria, using bacteriophages could eliminate antibiotics' damaging side effects.

Current Uses
Bacteriophages are not yet used as antibiotics because the immune system destroys the viruses before they can infect the pathogenic bacteria. Scientists are still researching ways to use bacteriophages effectively.

Bacteriophages are currently used to diagnose bacterial infections quickly. Diagnoses can be made by injecting many different types of bacteriophages into a patient. Blood tests then indicate which virus was able to reproduce, in turn determining the type of bacterial infection the patient has. Scientists are also able to use bacteriophages in the same way to detect bacterial contamination of food and water supplies. Perhaps one day in the future, your doctor will give you a helpful virus instead of an antibiotic to fight the harmful bacteria that make you sick!

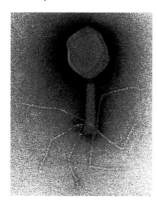

◀ *Some bacteriophages look more like machines than living organisms.*

Going Further
▶ Bacteriophages aren't always so helpful. Sometimes they can do more harm than good. Can you think of ways bacteriophages can cause trouble for humans ?

43

Answer to Going Further

Bacteriaphages can kill beneficial bacteria such as bacteria that help make food (lactic-acid bacteria), bacteria that supply plants with nitrogen (nitrogen-fixing bacteria), and bacteria that make medicine.

HEALTH WATCH
Helpful Viruses

Background
Bacteriophages were described first by F. W. Twort in 1915. It was immediately hoped that the bacteriophages could be used to combat pathogenic bacteria, but no one has discovered a way to do that successfully.

In the 1920s, bacteriophages were used to identify particular strains of bacteria. In the 1940s, Max Delbruck, Salvadore Luria, and other scientists began to discover the intricacies of bacteriophage reproduction. This precipitated the study of bacteriophage genetics. Scientists are now investigating how to use viruses as the vehicles to carry replacement copies of genes used for gene therapy.

Teaching Strategy
Have students research the structure of a bacteriophage. They should draw a bacteriophage and label its parts. They should then write a sentence or two describing the function of each part.

Chapter Organizer

CHAPTER ORGANIZATION	TIME MINUTES	OBJECTIVES	LABS, INVESTIGATIONS, AND DEMONSTRATIONS
Chapter Opener pp. 44–45	45	National Standards: SAI 1, SPSP 1, LS 1c, 1f, 4b	**Start-Up Activity,** A Microscopic World, p. 45
Section 1 Protists	90	▶ Describe the characteristics of protists. ▶ Name the three groups of protists, and give examples of each. ▶ Explain how protists reproduce. UCP 5, SPSP 1, LS 1b, 1c, 1f, 2a, 3a, 4b, 4c, 5a; Labs UCP 2	**Demonstration,** Algin and Carrageenan, p. 49 in ATE
Section 2 Fungi	90	▶ Describe the characteristics of fungi. ▶ Distinguish between the four main groups of fungi. ▶ Describe how fungi can be helpful or harmful. ▶ Define *lichen*. UCP 5, SAI 1, SPSP 5, HNS 1, LS 1b–1d, 1f, 2a, 4b, 4c; Labs SAI 1	**QuickLab,** Moldy Bread, p. 59 **QuickLab,** Observe a Mushroom, p. 61 **Skill Builder,** There's a Fungus Among Us! p. 64 **Datasheets for LabBook,** There's a Fungus Among Us! **Labs You Can Eat,** Knot Your Average Yeast Lab **Whiz-Bang Demonstrations,** Unleash the Yeast! **Long-Term Projects & Research Ideas,** Algae for All!

See page **T23** *for a complete correlation of this book with the*

NATIONAL SCIENCE EDUCATION STANDARDS.

TECHNOLOGY RESOURCES

 Guided Reading Audio CD
English or Spanish, Chapter 3

 One-Stop Planner CD-ROM with Test Generator

 CNN. **Eye on the Environment,** The Fire Ants and the Fungus, Segment 18

43A Chapter 3 • Protists and Fungi

CLASSROOM WORKSHEETS, TRANSPARENCIES, AND RESOURCES	SCIENCE INTEGRATION AND CONNECTIONS	REVIEW AND ASSESSMENT
Directed Reading Worksheet **Science Puzzlers, Twisters & Teasers**	**Across the Sciences,** It's Alive! p. 71	
Directed Reading Worksheet, Section 1 **Transparency 41,** *Euglena* **Transparency 121,** The Geologic Time Scale **Transparency 41,** *Paramecium* **Transparency 42,** The Life Cycle of *Plasmodium vivax* **Science Skills Worksheet,** Organizing Your Research **Math Skills for Science Worksheet,** A Shortcut for Multiplying Large Numbers **Critical Thinking Worksheet,** Protist Pop Culture **Reinforcement Worksheet,** Protists on Parade	**Multicultural Connection,** p. 48 in ATE **Cross-Disciplinary Focus,** p. 48 in ATE **Chemistry Connection,** p. 50 **Connect to Earth Science,** p. 50 in ATE **Geology Connection,** p. 53 **Connect to Earth Science,** p. 53 in ATE **Math and More,** Multiplying Protists, p. 55 in ATE	**Self-Check,** p. 48 **Section Review,** p. 51 **Self-Check,** p. 54 **Homework,** p. 55 in ATE **Section Review,** p. 56 **Quiz,** p. 56 in ATE **Alternative Assessment,** p. 56 in ATE
Directed Reading Worksheet, Section 2 **Reinforcement Worksheet,** An Ode to a Fungus	**Multicultural Connection,** p. 59 in ATE **MathBreak,** Multiplying Yeasts, p. 60 **Math and More,** The Price of Truffles, p. 60 in ATE **Apply,** p. 61 **Cross-Disciplinary Focus,** p. 62 in ATE **Real-World Connection,** p. 62 in ATE **Science, Technology, and Society:** Moldy Bandages, p. 70 **Across the Sciences,** It's Alive, p. 71	**Self-Check,** p. 58 **Self-Check,** p. 59 **Section Review,** p. 63 **Quiz,** p. 63 in ATE **Alternative Assessment,** p. 63 in ATE

internet connect

 Holt, Rinehart and Winston On-line Resources

go.hrw.com

For worksheets and other teaching aids related to this chapter, visit the HRW Web site and type in the keyword: **HSTPRO**

 National Science Teachers Association

www.scilinks.org

Encourage students to use the *sci*LINKS numbers listed in the internet connect boxes to access information and resources on the **NSTA** Web site.

END-OF-CHAPTER REVIEW AND ASSESSMENT

Chapter Review in Study Guide
Vocabulary and Notes in Study Guide
Chapter Tests with Performance-Based Assessment, Chapter 3 Test
Chapter Tests with Performance-Based Assessment, Performance-Based Assessment 3
Concept Mapping Transparency 11

Chapter Resources & Worksheets

Visual Resources

TEACHING TRANSPARENCIES

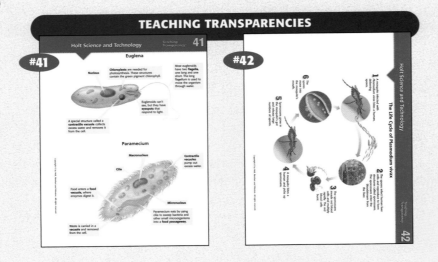

#41

Holt Science and Technology — Teaching Transparency 41

Euglena

Paramecium

#42

Holt Science and Technology

The Life Cycle of Plasmodium Vivax

TEACHING TRANSPARENCIES

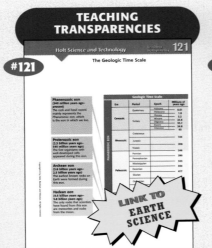

#121

Holt Science and Technology — Teaching Transparency 121

The Geologic Time Scale

Phanerozoic eon

Proterozoic eon

Archean eon

Hadean eon

LINK TO EARTH SCIENCE

CONCEPT MAPPING TRANSPARENCY

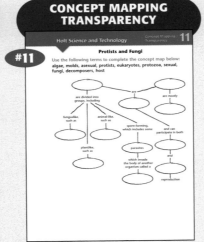

#11

Holt Science and Technology — Concept Mapping Transparency 11

Protists and Fungi

Use the following terms to complete the concept map below: algae, molds, asexual, protists, eukaryotes, protozoa, sexual, fungi, decomposers, host

---

Meeting Individual Needs

DIRECTED READING

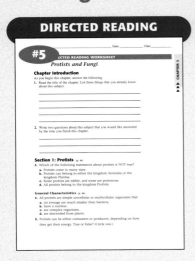

#5

DIRECTED READING WORKSHEET

Protists and Fungi

CHAPTER 3

Chapter Introduction

As you begin this chapter, answer the following.

1. Read the title of the chapter. List three things that you already know about this subject.

2. Write two questions about this subject that you would like answered by the time you finish this chapter.

Section 1: Protists

3. Which of the following statements about protists is NOT true?
   a. Protists come in many sizes.
   b. Protists can belong to either the kingdom Animalia or the kingdom Plantae.
   c. Some protists are edible, and some are poisonous.
   d. All protists belong to the kingdom Protista.

General Characteristics

4. All protists are simple unicellular or multicellular organisms that
   a. on average are much smaller than bacteria.
   b. have a nucleus.
   c. are complex organisms.
   d. are descended from plants.

5. Protists can be either consumers or producers, depending on how they get their energy. True or False? (Circle one.)

REINFORCEMENT & VOCABULARY REVIEW

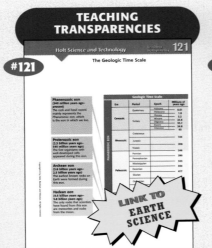

#5

REINFORCEMENT WORKSHEET

Protists on Parade

Complete the table below after you finish reading Chapter 11, Section 1.

Term	Definition	Example
Parasite		"late blight"
	makes its own food, usually by photosynthesis	
Consumer		scrambled egg slime mold
Flagella		
Cilia		
Pseudopodia		amoeba
Spore-forming		
	reproduction in which two organisms join together and exchange genetic material	
Fission		euglena

A Moldy Puzzle

#5

VOCABULARY REVIEW WORKSHEET

A Moldy Puzzle

After you finish reading Chapter 11, give this puzzle a try!
Fill in the blanks with the appropriate terms, and then use the terms to complete the puzzle on the next page.

1. An organism that invades the body of another organism is a _____.
2. _____ are fungi that do not fit into other standard groups of fungi.
3. _____ are animal-like protists that are single-celled consumers.
4. Plantlike protists that convert the sun's energy into food through photosynthesis are called _____.
5. The organism that is harmed by a parasite is called the _____.
6. A _____ is made of a fungus and an alga that grow intertwined.
7. Any eukaryotic organism that is part of the kingdom Protista is called a _____.
8. _____ like fungi include black bread mold.
9. Amoebas use _____ or "false feet," to move around.
10. _____ are chains of cells that make up multicellular fungi.
11. The group of fungi that includes umbrella-shaped mushrooms and puffballs is called _____.
12. A protist that obtains its food from dead organic matter or from the body of another organism is called a _____ protist.
13. A _____ is a small reproductive cell protected by a thick cell wall.
14. Fungi that reproduce by spores that develop in an ascus are called _____.
15. Kingdom _____ includes complex, multicellular organisms that obtain food by breaking down other substances in their surroundings and absorbing the nutrients.
16. The major part of a multicellular fungus is a twisted mass of hyphae that have grown together called a _____.
17. A shapeless, fuzzy fungus is a _____.
18. _____ are single-celled, microscopic, photosynthetic organisms that float near the surface of the ocean.

SCIENCE PUZZLERS, TWISTERS & TEASERS

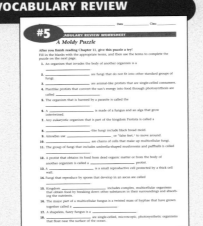

#5

SCIENCE PUZZLERS, TWISTERS & TEASERS

Protists and Fungi

Riddles

1. Try to solve the following riddles based on what you have learned about protists and fungi.
   a. I sometimes have structures that look like antennae, but I am not an alien.
      I like to eat, but then again I don't have to.
      I'm most comfortable among the dead.
      What am I?
   b. I can convert the sun's energy into food, but I am not a plant.
      Even if you are not mad, I make you see red.
      I can poison you without ever touching you.
      What am I?
   c. You have seen me on a pizza, taken me when you are sick, cleaned me off your shower wall, and I've made you scratch your feet.
      What am I?

The Scientist

2. You are conducting a study on animal-like protists. At the beginning of the study there are 4 amoebas, but 4 die, so 12 more are purchased. Halfway through the study, 3 disappear and 8 more are purchased. Three-quarters of the way through the study, 6 of the amoebas split in half. At the end of the project, the scientist looks in the mirror and lets out a sigh of relief.
   a. What color are the scientist's eyes? _____
   b. How do you know? _____

Chapter 3 • Protists and Fungi

Review & Assessment

STUDY GUIDE

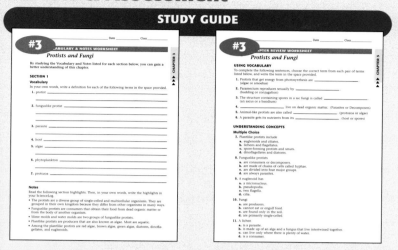

#3 VOCABULARY & NOTES WORKSHEET
Protists and Fungi

By studying the Vocabulary and Notes listed for each section below, you can gain a better understanding of this chapter.

SECTION 1

Vocabulary
In your own words, write a definition for each of the following terms in the space provided.
1. protist
2. funguslike protist
3. parasite
4. host
5. algae
6. phytoplankton
7. protozoa

Notes
Read the following section highlights. Then, in your own words, write the highlights in your ScienceLog.
• The protists are a diverse group of single-celled and multicellular organisms. They are grouped in their own kingdom because they differ from other organisms in many ways.
• Funguslike protists are consumers that obtain their food from dead organic matter or from the body of another organism.
• Slime molds and water molds are two groups of funguslike protists.
• Plantlike protists are producers that are also known as algae. Most are aquatic.
• Among the plantlike protists are red algae, brown algae, green algae, diatoms, dinoflagellates, and euglenoids.

#3 CHAPTER REVIEW WORKSHEET
Protists and Fungi

USING VOCABULARY
To complete the following sentences, choose the correct term from each pair of terms listed below, and write the term in the space provided.
1. Protists that get energy from photosynthesis are _____ (algae or amoebas)
2. Paramecium reproduces sexually by _____ (budding or conjugation)
3. The structure containing spores in a sac fungi is called _____ (an ascus or a basidium)
4. _____ live on dead organic matter. (Parasites or Decomposers)
5. Animal-like protists are also called _____ (protozoa or algae)
6. A parasite gets its nutrients from its _____ (host or spores)

UNDERSTANDING CONCEPTS

Multiple Choice
7. Plantlike protists include
a. euglenoids and ciliates.
b. lichens and flagellates.
c. spore-forming protists and smuts.
d. dinoflagellates and diatoms.
8. Funguslike protists
a. are consumers or decomposers.
b. are made of chains of cells called hyphae.
c. are divided into four major groups.
d. are always parasites.
9. A euglenoid has
a. a micronucleus.
b. pseudopodia.
c. two flagella.
d. cilia.
10. Fungi
a. are producers.
b. cannot eat or engulf food.
c. are found only in the soil.
d. are primarily single-celled.
11. A lichen
a. is a parasite.
b. is made up of an alga and a fungus that live intertwined together.
c. can live only where there is plenty of water.
d. is a consumer.

CHAPTER TESTS WITH PERFORMANCE-BASED ASSESSMENT

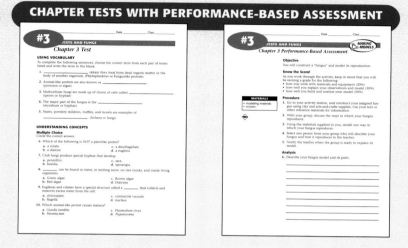

#3 PROTISTS AND FUNGI
Chapter 3 Test

USING VOCABULARY
To complete the following sentences, choose the correct term from each pair of terms listed below and write the term in the blank.
1. _____ obtain their food from dead organic matter or the body of another organism. (Phytoplankton or Funguslike protists)
2. Animal-like protists are also known as _____ (protozoa or algae)
3. Multicellular fungi are made up of chains of cells called _____ (spores or hyphae)
4. The major part of the fungus is the _____ (mycelium or hyphae)
5. Yeasts, powdery mildews, truffles, and morels are examples of _____ (lichens or fungi)

UNDERSTANDING CONCEPTS

Multiple Choice
Circle the correct answer.
6. Which of the following is NOT a plantlike protist?
a. a ciliate
b. a diatom
c. a dinoflagellate
d. a euglena
7. Club fungi produce special hyphae that develop
a. penicillin.
b. basidia.
c. sacs.
d. sporangia.
8. _____ can be found in water, in melting snow, on tree trunks, and inside living organisms.
a. Green algae
b. Red algae
c. Brown algae
d. Diatoms
9. Euglenas and ciliates have a special structure called a _____ that collects and removes excess water from the cell.
a. chloroplast
b. flagella
c. contractile vacuole
d. nucleus
10. Which animal-like protist causes malaria?
a. *Giardia lamblia*
b. *Paramecium*
c. *Plasmodium vivax*
d. *Trypanosoma*

#3 PROTISTS AND FUNGI
Chapter 3 Performance-Based Assessment — MAKING MODELS

Objective
You will construct a "fungus" and model its reproduction.

Know the Score!
As you work through the activity, keep in mind that you will be earning a grade for the following:
• how you work with materials and equipment (20%)
• how well you explain your observations and model (30%)
• how well you build and analyze your model (50%)

MATERIALS
• modeling materials
• scissors

Procedure
1. Go to your activity station, and construct your assigned fungus using clay and arts-and-crafts supplies. Use your text or other reference materials for information.
2. With your group, discuss the ways in which your fungus reproduces.
3. Using the materials supplied to you, model one way in which your fungus reproduces.
4. Select one person from your group who will describe your fungus and how it reproduces to the teacher.
5. Notify the teacher when the group is ready to explain its model.

Analysis
6. Describe your fungus model and its parts.

Lab Worksheets

LABS YOU CAN EAT

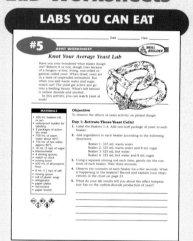

#5 STUDENT WORKSHEET — SKILL BUILDER
Knot Your Average Yeast Lab

Have you ever wondered what makes dough rise? Believe it or not, dough rises because of a fungus—a tiny, living, one-celled organism called yeast. When dried, yeast are in a state of suspended animation. But when you add warm water and sugar, watch out! The yeast get active and go into a feeding frenzy. What's left behind is carbon dioxide and alcohol.
In this activity, you can watch yeast at work!

MATERIALS
• 500 mL beakers (4) or jars
• waterproof marker for labeling
• 2 packages of active dry yeast
• 750 mL of warm water about 40°C
• 16 mL (1 tsp) of sugar
• thermometer
• 4 stirring spoons
• watch or clock
• mixing bowl
• 650 mL of all-purpose flour
• 8 mL (1½ tsp) of salt
• mixing spoon
• roll of plastic wrap
• refrigerator
• paper plates
• microwave
• paper towels

Objective
To observe the effects of yeast activity on pretzel dough

Day 1: Activate Those Yeast Cells!
1. Label the beakers 1–4. Add one-half package of yeast to each beaker.
2. Add ingredients to each beaker according to the following directions:
Beaker 1: 325 mL warm water
Beaker 2: 325 mL warm water and 8 mL sugar
Beaker 3: 325 mL cold water
Beaker 4: 325 mL hot water and 8 mL sugar
3. Using a separate stirring rod each time, gently stir the contents of each beaker. Wait three minutes.
4. Observe the contents of each beaker for a few seconds. What is happening in the beakers? Record and explain your observations in the chart on page 23.
5. What do your lab results tell you about the effect temperature has on the carbon-dioxide production of yeast?

LONG-TERM PROJECTS & RESEARCH IDEAS

#3 STUDENT WORKSHEET — DESIGN YOUR OWN
Algae for All!

What do pond scum, red tide, and the green stuff growing in your aquarium have in common? They all are types of algae! As a matter of fact, algae are valuable members of the biosphere. Unfortunately, algae don't always receive the recognition they deserve. Did you know that half the world's organic material produced by photosynthesis come from algae? And humans use algae more than you might think. We use algae to make medicines as well as to treat sewage. Many cultures eat seaweed and other algae regularly. And though they thrive in watery places like lakes and ponds, these hardy creatures can live almost anywhere—from the Antarctic to the Sahara Desert!

USEFUL TERMS
dermatologist a doctor who specializes in the skin and its diseases
pathologist a doctor who studies the effects of diseases on the body

Algae Blooms
1. How do algae and detergent mix? Design an experiment to determine the effects of detergent in waste water on algae and other pond plants, such as *Elodea*. Repeat your experiment to test for the effects of fertilizers or acid rain. Present your findings in a scientific article.

More Long-Term Projects
2. Athlete's foot is an itchy infection caused by a fungus. Interview a dermatologist or pathologist about other human diseases that are caused by fungi or by protozoa. Create a brochure for patients with a fungal- or protozoan-related disease that explains the disease and its treatment.
3. You may not realize it, but you probably eat algae on a regular basis. No kidding! Carrageenan, mannitol, and agar are a few algae extracts commonly used in food products. Look for these extracts and other algae-related products on the labels of food and cosmetic containers. Identify and list products in your home that contain algae. Create a poster that highlights at least five of the products in your home that use or contain algae.

Research Idea
4. Sometimes an organ-transplant patient suffers from organ rejection, even though a new organ can save the patient's life! Why does the body do this? How can the drug cyclosporine, derived from a fungus, help the transplant operation to be more successful? Find out about cyclosporine, and research other medical uses of fungi. Write a newspaper article about your findings.

WHIZ-BANG DEMONSTRATIONS

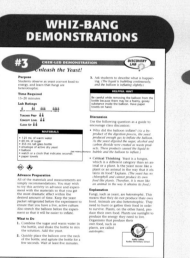

#3 TEACHER-LED DEMONSTRATION — DISCOVERY LAB
Unleash the Yeast!

Purpose
Students observe as yeast convert food to energy, and learn that fungi are heterotrophic.

Time Required
15–20 minutes

Lab Ratings
TEACHER PREP
CONCEPT LEVEL
CLEAN UP

MATERIALS
• 125 mL of warm water
• 60 mL of sugar
• 355 mL tall glass bottle
• envelope of active dry yeast
• balloon
• watch or a clock that indicates seconds
• paper towels

Advance Preparation
All of the materials and measurements are simply recommendations. You may want to try this activity in advance and experiment with the materials so that you get the most dramatic effect within the desired amount of time. Keep the yeast packet refrigerated before the experiment to ensure that you have a live, active culture. Also stretch the balloon before the experiment so that it will be easier to inflate.

What to Do
1. Combine the sugar and warm water in the bottle, and shake the bottle to mix the solution. Add the yeast.
2. Quickly place the balloon over the neck of the bottle, and agitate the bottle for a few seconds. Wait at least five minutes.
3. Ask students to describe what is happening. (*The liquid is bubbling continuously, and the balloon is inflating slightly.*)

HELPFUL HINT
Be careful while removing the balloon from the bottle because there may be a foamy, gooey substance inside the balloon. Have paper towels on hand.

Discussion
Use the following question as a guide to encourage class discussion:
• Why did the balloon inflate? (*As a by-product of the digestion process, the yeast produced enough gas to inflate the balloon. As the yeast digested the sugar, alcohol and carbon dioxide were created as waste products. These products caused the liquid to bubble and the balloon to inflate.*)
• **Critical Thinking** Yeast is a fungus, which is a different category than an animal or a plant. Is the yeast more like a plant or an animal in the way that it obtains its food? Explain. (*The yeast has no chlorophyll and cannot produce its own food like plants. Therefore, it is more like an animal in the way it obtains its food.*)

Explanation
Fungi, such as yeast, are heterotrophic. This means that they do not produce their own food. Animals are also heterotrophic. They need to hunt or gather their food in order to survive. Plants, on the other hand, produce their own food using sunlight to produce the energy they need to live. Organisms that produce their own food, such as plants, are called autotrophs.

DATASHEETS FOR LABBOOK

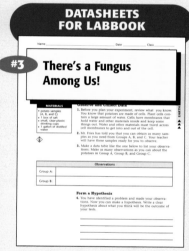

Name _____ Date _____ Class _____

#3 **There's a Fungus Among Us!**

MATERIALS
• potato samples (A, B, and C)
• 1 box of salt
• small, clear-plastic drinking cups
• 1 gallon of distilled water

Observe and Collect Data
1. Before you plan your experiment, review what you know. You know that potatoes are made of cells. Plant cells contain a large amount of water. Cells have membranes that hold water and other materials inside and keep other things out. Water and other materials must travel across cell membranes to get into and out of the cell.
2. Mr. Fries has told you that you can obtain as many samples as you need from Groups A, B, and C. Your teacher will have three samples ready for you to observe.
3. Make a data table like the one below to list your observations. Make as many observations as you can about the potatoes in Group A, Group B, and Group C.

Observations	
Group A:	
Group B:	

Form a Hypothesis
4. You have identified a problem and made your observations. Now you can make a hypothesis. Write a clear hypothesis about what you think will be the outcome of your tests.

Applications & Extensions

CRITICAL THINKING & PROBLEM SOLVING

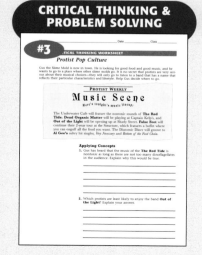

#3 CRITICAL THINKING WORKSHEET
Protist Pop Culture

Gus the Slime Mold is new in town. He is looking for good food and good music, and wants to go to a place where other slime molds go. It is no secret that protists are very serious about their musical choices—they will only go to listen to a band that has a name that reflects their particular characteristics and lifestyle. Help Gus decide where to go.

PROTIST WEEKLY

Music Scene
Here's tonight's music lineup:

The Underwater Cafe will feature the nontoxic sounds of **The Red Tide**. **Dead Organic Matter** will be opening up at Captain Kelp's, and **Out of the Light** will be opening up at Shady Street. **False Feet** will continue their 2-year tour at the Structure, which features a buffet where you can engulf all the food you want. The Diatomic Diner will groove to **Al Gee's** sultry hit singles, *Very Necessary* and *Bottom of the Food Chain*.

Applying Concepts
1. Gus has heard that the music of the **The Red Tide** is nontoxic as long as there are not too many dinoflagellates in the audience. Explain why this would be true.

2. Which protists are least likely to enjoy the band **Out of the Light**? Explain your answer.

EYE ON THE ENVIRONMENT

#18 Science in the News: Critical Thinking Worksheets
Segment 18

The Fire Ants and the Fungus

1. Name two ways that exotic species may be accidentally introduced to a new ecosystem.

2. Why is the source country often the same for both the pest and the possible biological solution?

3. Why did the researchers patent the information they learned about the fungus?

4. Describe two trade-offs between the harm the fire pest-control agents on the environment and human...

Protists

▶ Protists, Protists, Everywhere

Most protists are aquatic. Some live in marine environments; others live in fresh water or in the water that surrounds soil particles. Still other protists live in the body fluids of other organisms.

▶ Products from Algae

The cell walls of red algae contain a mucous substance that gives the algae a slippery texture. Agar, which is derived from this substance, has the consistency of gelatin and is used worldwide as a culture medium for growing bacteria.

- Giant kelp and other brown algae are the source of algin, which has hundreds of uses. For example, it is used as a thickening and stabilizing agent in ice cream, milkshakes, pie fillings, and weight-control drinks; as an additive in paper and a coating on frozen food packages; as a smoothing agent in lotions and creams; and as an ingredient in latex paints and adhesives.

IS THAT A FACT!

- ➤ During the Irish potato famine of 1846, people ate a red alga called dulse as a substitute for potatoes. Another red alga, commonly called purple laver or nori, is used extensively in food today, especially in Asia.

▶ Slime Molds

There are two types of slime molds: cellular slime molds (about 70 species in the phylum Acrasiomycota) and plasmodial or acellular slime molds (roughly 500 species in the phylum Myxomycota).

- Plasmodial slime molds are named for their slimy, often large and colorful plasmodia. A plasmodium is the feeding phase of the slime mold, and it engulfs bacteria, yeast, and bits of organic matter in its path. A plasmodium can flow around obstacles and will even flow through the meshwork of a piece of cloth.

- ➤ Many students who have kept fish in an aquarium have seen the common water mold *Saprolegnia,* which forms a fuzzy white mass as it grows over the surface of a dead or injured fish.

▶ Diatoms

The word *diatom* comes from the Greek *diatomos,* which means "cut in two." It refers to the glassy, two-part shells (called frustules) that enclose these single-celled organisms.

- The frustules of diatoms have complex and strikingly beautiful markings that are different for each species and are therefore important in diatom identification.

- One liter of sea water may contain almost a million diatoms.

IS THAT A FACT!

- ➤ A harmless species of symbiotic amoeba called *Entamoeba gingivalis* lives in the mouths of many people, where it feeds on loose cells and organic debris.

▶ The Need to Conjugate

Laboratory experiments have shown that some species of *Paramecium* must conjugate periodically to survive. If they are not allowed to conjugate, these *Paramecium* have the capacity for only a limited number of asexual divisions (about 350) before they die.

SECTION 2

Fungi

▶ Fungi Functions

About 50,000 species of fungi have been identified. Fungi are extremely important as decomposers; they break down complex organic material to simple organic compounds and inorganic molecules, making carbon, nitrogen, phosphorus, and other essential elements available to living things.

- Fungi are also essential in the making of bread, cheeses, wine, beer, and soy sauce; in the production of many antibiotics; and as research organisms for geneticists, biochemists, cytologists, and microbiologists.

- On the other hand, fungi are the major cause of plant diseases. There are at least 5,000 different kinds of fungi that attack crops, garden plants, and wild plants. Some fungi also cause disease in animals, such as ringworm.

▶ Sac Fungi

There are more than 30,000 known species of sac fungi (class Ascomycota), about 500 of which are single-celled yeasts.

- Morels and truffles are multicellular sac fungi. For hundreds of years, the only way to enjoy truffles, which grow underground on the roots of oak and hazelnut trees, was to unearth wild ones with the help of specially trained "truffle-hunting" pigs and dogs. Truffles are now cultivated commercially in France, but they remain a very expensive delicacy.

▶ True Mushrooms

Club fungi include the true mushrooms. There are about 25,000 known species of club fungi.

IS THAT A FACT!

- One mycelium of the club fungus *Armillaria bulbosa,* is among of the world's oldest and largest organisms. Found in a Michigan forest, this particular mycelium is estimated to be more than 1,500 years old, to weigh more than 10,000 kg, and to cover an area of 150,000 m^2.

▶ A Predatory Fungus

One of the imperfect fungi *(Arthrobotrys dactyloides)* preys on tiny roundworms (nematodes) in soil. The filaments of the fungi produce minute "nooses" that swell rapidly when nematodes try to crawl through them. The nooses hold the nematodes tightly while hyphae grow into their bodies and kill them.

▶ Lichens

Approximately 15,000 species of lichens have been identified. Lichens are a symbiotic association between a fungus—in most cases a sac fungus, or ascomycete—and a species of photosynthetic algae or cyanobacteria.

- In severe growing conditions, lichens grow extremely slowly; even small lichens may be hundreds or even thousands of years old.

IS THAT A FACT!

- Lichens are one of the most important foods eaten by caribou. During winter on the tundra, when other foods are not available, caribou survive almost exclusively on a diet of lichens.

> **For background information about teaching strategies and issues, refer to the *Professional Reference for Teachers.***

Protists and Fungi

Protists and Fungi

Pre-Reading Questions

Students may not know the answers to these questions before reading the chapter, so accept any reasonable response.

Suggested Answers

1. Seaweed is a type of alga.

2. A fungus is a eukaryotic consumer.

3. Mushrooms do not have roots. They grow from hyphae.

Sections

Pre-Reading Questions

1. What is seaweed?
2. What is a fungus?
3. Do mushrooms have roots?

HARDWORKING MUSHROOMS!

The kingdoms Protists and Fungi (FUHN JIE) contain many fascinating and beneficial organisms. Protists make most of Earth's oxygen. Many trees and other plants need the help of fungi to get nutrients from the soil. The mushrooms on these pages are fungi that help break down dead plant matter on the forest floor. This process helps recycle the nutrients of the forest. In this chapter, you will learn more about protists and fungi.

---

internet connect

 HRW On-line Resources

go.hrw.com
For worksheets and other teaching aids, visit the HRW Web site and type in the keyword: **HSTPRO**

www.scilinks.com
Use the *sci*LINKS numbers at the end of each chapter for additional resources on the **NSTA** Web site.

www.si.edu/hrw
Visit the Smithsonian Institution Web site for related on-line resources.

www.cnnfyi.com
Visit the CNN Web site for current events coverage and classroom resources.

START-UP Activity

A MICROSCOPIC WORLD

In this activity, you will observe some common protists in pond water or in a solution called a *hay infusion*.

Procedure

1. Using a **plastic eyedropper,** place one drop of **pond water** or **hay infusion** onto a **microscope slide.**

2. Add one drop of **ProtoSlo™** to the drop on the slide.

3. Add a **plastic coverslip** by putting one edge on the slide and then slowly lowering it over the drop to prevent air bubbles.

4. Observe the slide under low power of a **microscope.** Once you've located an organism, try high power for a closer look.

5. In your ScienceLog, sketch the organisms you see under high power.

Analysis

6. How many different organisms do you see?

7. Are the organisms alive? Support your answer with evidence.

8. How many cells does each organism appear to have?

45

START-UP Activity

A MICROSCOPIC WORLD

MATERIALS
FOR EACH STUDENT:
• plastic dropper
• pond water or hay infusion
• ProtoSlo™
• microscope slide
• plastic coverslip
• microscope

Safety Caution

Tell students not to taste the solution. Care should be taken handling microscope slides and coverslips. check for known mold or fungi reactions among students before conducting this lab. Have an eyewash available in the even of a splash to the eye. Instruct students to wipe up all spills immediately.

Teacher's Notes

Pond water may contain plant-like protists, such as *Spirogyra* and *Volvox,* and animal-like protists, such as *Stentor, Vorticella, Euglena, Paramecium,* and amoebas. Students may also see nematode worms and small fast-moving multicellular animals called rotifers.

Answers to START-UP Activity

6. Answers will vary.

7. The organisms are moving, which suggests that they are alive.

8. Answers may vary, but most, if not all, are single-celled organisms.

Protists

This section introduces students to organisms that make up the kingdom Protista. Students learn that protists are commonly grouped into three categories—funguslike protists, plantlike protists, and animal-like protists —based on the way in which they obtain nutrients. Students are introduced to representatives of each protist group and also learn about reproduction in protists.

Bellringer

Ask students to imagine that an organism needs to transport itself without the use of arms, fins, wings or legs. How would the organism do it? What environments could the organism live in?

Have students write their answers in their ScienceLog. Encourage students to illustrate their answers whenever possible. (Accept all reasonable responses. Tell students that many protists live in water and move with the help of flagella and cilia.)

Directed Reading Worksheet Section 1

Terms to Learn

protist	host
funguslike protist	algae
	phytoplankton
parasite	protozoa

What You'll Do

◆ Describe the characteristics of protists.
◆ Name the three groups of protists, and give examples of each.
◆ Explain how protists reproduce.

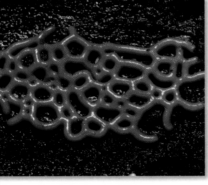

Figure 1 *Protists have many different shapes and sizes.*

Pretzel slime mold

Protists

Some are so tiny they cannot be seen without a microscope, and others grow many meters long. Some are poisonous, and others provide food. Some are like plants. Some are like animals. And some are nothing like plants or animals. Despite their differences, all of these organisms are related. What are they? They are all members of the kingdom Protista and are called **protists.** Look at **Figure 1** to see some of the variety of protists.

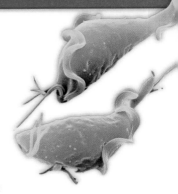

Zooflagellate

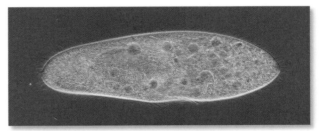

Paramecium

Ulva

General Characteristics

All protists are *eukaryotic*. That means their cells have a nucleus. Most protists are single-celled organisms, but some are multicellular. Scientists generally agree that the more complex eukaryotic organisms—plants, animals, and fungi—all originated from primitive protists.

Some protists are *producers*. Like plants, they get their energy from the sun through *photosynthesis*. Others are *consumers*. They cannot obtain energy from sunlight and must get food from their environment. Protists are often classified by the way they obtain energy. This method groups these organisms into funguslike protists, plantlike protists, or animal-like protists.

IS THAT A FACT!

The glassy "shells" of many kinds of diatoms are perforated by tiny holes. Although most diatoms contain chlorophyll and carry out photosynthesis, many also are able to absorb minerals through these perforations.

Funguslike Protists

A fungus is an organism that obtains its food from dead organic matter or from the body of another organism. You will learn more about fungi in the next section. The protists that get food this way are called **funguslike protists.** The funguslike protists are consumers that secrete digestive juices into the food source and then absorb the digested nutrients. These protists also reproduce like fungi. Two types of funguslike protists will be discussed in this chapter—slime molds and water molds.

It's Slime! *Slime molds* are thin masses of living matter. They look like colorful, shapeless globs of slime. Many slime molds live as single-celled organisms. But during times of environmental stress, these single organisms come together to form a group of cells with many nuclei and a single cytoplasm. Slime molds live in cool, shady, moist places in the woods and in fresh water. **Figure 2** shows a slime mold growing over a log.

Slime molds eat bacteria, yeast, and small bits of decaying plant and animal matter. They surround food particles and digest them. As long as food and water are available, a slime mold will continue to grow. It may cover an area more than 1 m across!

When growth conditions are unfavorable, a slime mold develops stalklike structures with rounded knobs at the top. You can see this in **Figure 3.** The knobs contain *spores*. The spores can survive for a long time without water or nutrients. When conditions improve, the spores will develop into new slime molds.

Figure 2 *Slime molds, like this scrambled egg slime mold, are consumers.*

Figure 3 *The spore-containing knobs of a slime mold are called sporangia.*

47

WEIRD SCIENCE

At the 1933 Chicago World's Fair, an exhibit of "hair growing on wood" was displayed in the Believe It or Not pavilion. Although the "hair" amazed many fair-goers, it was actually the clustered fruiting bodies of a slime mold.

1) Motivate

MAKING MODELS

Slime Mold For this activity, students should wear safety goggles, an apron, and protective gloves. They should not put any of the materials in their mouth. Food coloring is nontoxic but can stain skin and clothing.

Students can make a substance that behaves very much like a slime mold. Divide the class into groups of 4–5 students. Each group should have a 25 × 25 cm (9 × 9 in.) pan, cornstarch, food coloring, a small container of water, a measuring cup, and a tablespoon. Instruct the students to place 240 mL (1 cup) of cornstarch in the pan. Then have them add 15 mL (1 tbsp) of water at a time to the cornstarch until the mixture has the consistency of a slime that flows slowly. If they add too much water, have them add a little more cornstarch to get the right consistency. Then mix a few drops of food coloring into the mixture and allow students to explore moving the slime around in the pan. They can put small objects, such as paper clips, in the path of the slime and let the substance surround the objects.
Sheltered English

Prediction Guide Before students read about algae, ask them the following question:

Knowing that algae carry out photosynthesis, what substance would you expect to find in their cells? (the green pigment chlorophyll)

ACTIVITY

Writing **Concept Mapping**
Have students construct a concept map using the following words and phrases:

funguslike protists, water mold, slime mold, protists, bacteria, producers, consumers

Answers to Self-Check

No; some funguslike protists are parasites or consumers.

🌐 Multicultural CONNECTION

Native Hawaiians have historically used a species of brown algae, a type of *Sargassum,* to heal cuts caused by corals. The algae was chopped up and applied to cuts as a poultice.

Other seaweeds contain compounds that may have medicinal value as well. About 20 different seaweeds are used in preparations for treating diseases, including intestinal parasite infections and cancer.

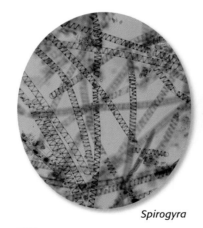

Figure 4 *Parasitic water molds attack various organisms, including fish.*

Moldy Water? Another type of funguslike protist is the *water mold.* Most water molds are small, single-celled organisms. Water molds live in water, moist soil, or other organisms.

Some water molds are *decomposers* and eat dead organic matter. But many water molds are parasites. **Parasites** invade the body of another organism to obtain the nutrients they need. The organism a parasite invades is called a **host.** Hosts can be living plants, animals, algae, or fungi. A parasitic water mold is shown in **Figure 4.**

Some parasitic water molds cause diseases. A water mold causes "late blight" of potatoes, the disease that led to the Great Potato Famine in Ireland from 1845—1852. Another water mold attacks grapes and threatened the French wine industry in the late 1800s. These protists still endanger crops today, but fortunately methods now exist to control them.

Plantlike Protists: Algae

A second group of protists are producers. Like plants, they use the sun's energy to make food through photosynthesis. These plantlike protists are also known as **algae** (AL JEE). All algae (singular, *alga*) have the green pigment chlorophyll, which is used for photosynthesis. But most algae also have other pigments that give them a specific color. Almost all algae live in water. You can see some examples of algae in **Figure 5.**

Some algae are multicellular. These algae generally live in shallow water along the shore. You may know these as *seaweed* or *kelp.* Some of these algae can grow to many meters in length.

> ✓ **Self-Check**
>
> Are all funguslike protists decomposers? Explain. *(See page 168 to check your answers.)*

Figure 5 *Algae range in size from giant seaweeds to single-celled organisms.*

Spirogyra

Kelp

48

Art Anna Atkins (1799–1871) was an amateur botanist. Atkins specialized in making photographic blueprints called cyanotypes, or sunprints. Over a 10-year period, Atkins created hundreds of cyanotypes of algae from the British Isles. Her book, *British Algae: Cyanotype Impressions,* was published in 1843 and is thought to be the first book illustrated with photographic images. Interested students may wish to create their own cyanotypes of algae (or leaves, fern fronds, or other plant parts if algae are not available) to accompany their reports. Cyanotype paper can be obtained from biological supply houses.

Single-celled algae cannot be seen without a microscope. They usually float near the water's surface. The single-celled algae make up **phytoplankton** (FITE oh PLANK tuhn). Phytoplankton are producers that provide food for most other water-dwelling organisms. They also produce most of the world's oxygen.

The plantlike protists are divided into phyla based on their color and cell structure. We will discuss six of the phyla here: red algae, brown algae, green algae, diatoms, dinoflagellates, and euglenoids.

Red Algae Most of the world's seaweeds are red algae. They contain chlorophyll and a red pigment that gives them their color. These multicellular protists live mainly in tropical marine waters, attached to rocks or other algae. Their red pigment allows them to absorb the light that filters deep into the clear water of the Tropics. Red algae can grow as much as 260 m below the surface of the water but are usually less than 1 m in length.

Brown Algae Most of the seaweeds found in cool climates are brown algae. They attach to rocks or form large floating beds in ocean waters. Brown algae have chlorophyll and a yellow-brown pigment. Many are very large—some grow 60 m in just one growing season! The tops of these gigantic algae are exposed to sunlight. The food made here by photosynthesis is transported to the parts of the algae that are too deep in the water to receive sunlight. An example of a brown alga can be seen in **Figure 6.**

Figure 6 Laminaria *is a brown alga.*

Green Algae The green algae are the most diverse group of plantlike protists. They are green because chlorophyll is the main pigment they contain. Most live in water or moist soil, but others are found in melting snow, on tree trunks, and even inside other organisms.

Many green algae are single-celled, microscopic organisms. Others are multicellular. These species may grow up to 8 m long. Individual cells of some species of green algae live in groups called colonies. **Figure 7** shows colonies of *Volvox.*

Figure 7 Volvox *is a green alga that grows in round colonies.*

49

IS THAT A FACT!

Giant kelp, a type of brown algae, is anchored to the ocean bottom. It has air bladders in its stems that help keep the stem floating upward, toward the water's surface, where the sunlight is the strongest.

MEETING INDIVIDUAL NEEDS

Writing **Learners Having Difficulty** To help students organize information about the three types of algae described on page 49, have them create a chart with three columns, one for each type of algae. Each column should have three rows: one describing the size of the algae, one describing where the algae can be found, and one describing how the algae get their food. Then have students reread page 49 and record information about the different kinds of algae in the appropriate spaces on their charts. **Sheltered English**

DEMONSTRATION

Algin and Carrageenan
Display the following products (or product packages) on a table in the classroom:

    ice cream, salad dressing, jelly beans, chocolate milk or eggnog, instant pudding, dry gravy mix or canned gravy, marshmallows, and a nondairy dessert topping

Tell students that two substances extracted from algae—algin and carrageenan—are common ingredients in many foods. Algin comes from giant kelp and other brown algae, while carrageenan is extracted from a red alga. Both are used as food thickeners, stabilizers, and emulsifiers. Have students read the ingredients on the packages and discuss how widespread the use of algin and carrageenan is in commercially prepared foods. **Sheltered English**

internetconnect

SCI LINKS
NSTA

TOPIC: Algae
GO TO: www.scilinks.org
sciLINKS NUMBER: HSTL255

READING STRATEGY

Mnemonics Tell students that the word *phytoplankton* comes from the Greek words *phyto,* which means "plant," and *planktos,* which means "wandering." (Remind students that phytoplankton are now classified as protists, not as plants.) Tell students that plankton also contains tiny animals that as a group are referred to as *zooplankton;* the Greek root word *zoion* means "animal." An easy way to remember the difference is that both *phytoplankton* and *plant* begin with the letter *p,* while *zooplankton* contains the word *zoo,* which is a place where animals are kept.

CONNECT TO
EARTH SCIENCE

Diatomaceous earth is a sedimentary rock composed mainly of the glass shells of diatoms. It has many commercial uses. It is used as a filter for oils, chemicals, and even some beverages; as an abrasive in many kinds of polishes; and in the manufacture of ceramics.

Diatomaceous earth is spread in gardens to kill slugs. When these 10-million-year-old diatoms are crushed, they become as sharp as glass and slice the slugs on contact.

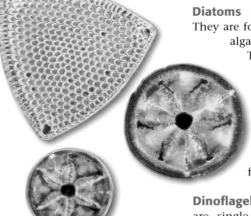

Figure 8 *Although most diatoms are free floating, many cling to plants, shellfish, sea turtles, and whales.*

Chemistry
CONNECTION

Some dinoflagellates give off light. A chemical reaction in the cells produces light that is similar to the light produced by fireflies. Water filled with these dinoflagellates glows like a twinkling neon light.

Figure 9 *Red tides occur throughout the world and are common in the Gulf of Mexico.*

Diatoms Diatoms (DIE e TAHMZ) are single-celled organisms. They are found in both salt water and fresh water. As with all algae, diatoms get their energy from photosynthesis. They make up a large percentage of phytoplankton. As you can see in **Figure 8,** many diatoms have unusual shapes. Their cell walls contain cellulose and silica, a rigid, glasslike substance. The cells are enclosed in a shell with two parts that fit neatly together. Piles of diatom shells deposited over millions of years form a fine, crumbly substance that is used in silver polish, toothpaste, filters, and insulation.

Dinoflagellates Most dinoflagellates (DIE noh FLAJ uh lits) are single-celled algae. They live primarily in salt water, although a few species live in fresh water, and some are even found in snow. Dinoflagellates have two whiplike strands called *flagella* (singular, *flagellum*). The beating of these flagella causes the cells to spin through the water. For this reason they are sometimes called spinning flagellates.

Most dinoflagellates get energy from photosynthesis, but a few are consumers, decomposers, or parasites. Some dinoflagellates are red and produce a strong poison. If these algae multiply rapidly, they can turn the water red, causing a dangerous condition known as *red tide.* When shellfish eat these algae, the poison is concentrated in their bodies. The shellfish are then toxic to humans and other vertebrates who eat them. A red tide is shown in **Figure 9.**

SCIENCE HUMOR

Knock, Knock!

Q: Who's there?

A: *Euglena.*

Q: *Euglena* who?

A: *Euglena* your room, or you're grounded!

Euglenoids Euglenoids (yoo GLEE NOYDZ) are single-celled protists that live primarily in fresh water. Most euglenoids have characteristics of both plants and animals. Like plants, they use photosynthesis. But when light is too low for photosynthesis, they can become consumers, like animals. Euglenoids can also move like animals. Flagella propel the organisms through the water. The structure of a euglenoid is shown in **Figure 10**.

Some euglenoids do not have chloroplasts for photosynthesis. These species either consume other small protists or absorb dissolved nutrients.

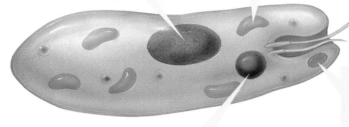

Nucleus

Chloroplasts are needed for photosynthesis. These structures contain the green pigment chlorophyll.

Most euglenoids have two **flagella**, one long and one short. The long flagellum is used to move the organism through water.

Euglenoids can't see, but they have **eyespots** that respond to light.

A special structure called a **contractile vacuole** collects excess water and removes it from the cell.

Figure 10 *Euglenoids have both plant and animal characteristics.*

SECTION REVIEW

1. How does a slime mold survive when food and water are limited?

2. Which plantlike protists move? How?

3. Add the following terms to the concept map at right: consumer, water mold, diatom, euglenoid.

4. Look at the picture of a euglenoid on this page. Which cell structures are plantlike? Which are animal-like?

5. **Analyzing Relationships** How do funguslike protists differ from plantlike protists?

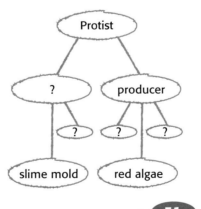

Teaching Transparency 41 *"Euglena"*

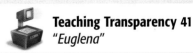

RETEACHING

Writing Have students read the description of a typical euglenoid on this page, and study **Figure 10.** Then ask students to imagine they are writing a letter to a friend describing a euglenoid and its animal-like and plantlike characteristics. Have students write their letters in their ScienceLog. **Sheltered English**

SCIENTISTS AT ODDS

Euglenoids pose a special problem for taxonomists because they exhibit characteristics of both plants and animals. Most euglenoids have plastids containing pigments similar to those in green algae, and they can carry on photosynthesis in sunlight. These are characteristics of algae. But euglenoids do not have cell walls, and they use flagella to propel themselves through water or anchor themselves in place. When a euglenoid is kept in the dark, it loses its green color and begins to feed like an animal. In sunlight it photosynthesizes just like a plant. For these reasons biologists used to classify euglenoids as protozoa in the phylum Zoomastigophora. Now they are classified as euglenoids in the phylum Euglenophyta.

▼ **Answers to Section Review**

1. Slime molds form spores when water and food are limited. These spores can survive poor growth conditions and will develop into new individuals when conditions improve.

2. Dinoflagellates and euglenoids move. Both organisms use flagella to move through the water.

3. Students might indicate a protist can be a consumer, such as a water mold or a slime mold. A protist can also be a producer, such as a diatom, a red algae, or a euglenoid.

4. In euglenoids, chloroplasts are plantlike. Eyespots and flagella are animal-like.

5. Funguslike protists are consumers. They get nutrients from other organisms. Plantlike protists are producers and can use photosynthesis.

PORTFOLIO

Tell students that the amoeba shown in **Figure 12** is engulfing prey. Ask students to make drawings in their ScienceLog that show all the stages in endocytosis—the amoeba's pseudopodia surrounding the prey, joining at their tips, and then forming a food vacuole around the engulfed organism. Then challenge students to create a similar series of drawings that illustrate the process of exocytosis (expelling a waste-filled vacuole) in an amoeba.

Sheltered English

ACTIVITY

Observing Live Amoebas

Obtain live amoebas from a biological supply house. Have students work in pairs to place an amoeba on a slide with a little water and observe it under a microscope. In their ScienceLog, have students describe how the amoeba moves. Students should also look for contractile vacuoles in their specimens.

Sheltered English

internet**connect**

SC*i*LINKS.
NSTA

TOPIC: Protozoa
GO TO: www.scilinks.org
*sci*LINKS NUMBER: HSTL260

Animal-Like Protists: Protozoa

The animal-like protists are single-celled consumers. These protists are also known as **protozoa**. Some are parasites. Many can move. Scientists do not agree on how to group protozoa, but they are often divided into four phyla: amoebalike protists, flagellates, ciliates, and spore-forming protists.

Amoebalike Protists An amoeba (uh MEE buh) is a soft, jellylike protozoan. Amoebas are found in both fresh and salt water, in soil, or as parasites in animals. Although an amoeba looks shapeless, it is actually a highly structured cell. Like euglenoids, amoebas have contractile vacuoles to get rid of excess water. Amoebas move with *pseudopodia* (soo doh POH dee uh). *Pseudopodia* means "false feet." You can see how an amoeba uses pseudopodia to move in **Figure 11.**

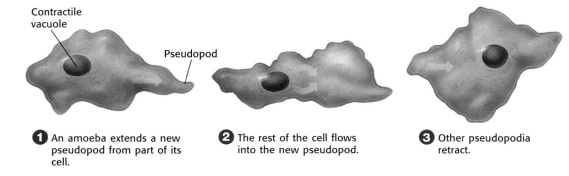

Contractile vacuole

Pseudopod

1 An amoeba extends a new pseudopod from part of its cell.

2 The rest of the cell flows into the new pseudopod.

3 Other pseudopodia retract.

Figure 11 *The shape of an amoeba changes constantly as new pseudopodia form.*

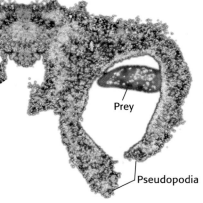

Prey

Pseudopodia

Figure 12 *An amoeba engulfs its prey with its pseudopodia.*

Feeding Amoebas Like slime molds, amoebas feed by engulfing food. An amoeba senses the presence of another single-celled organism and moves toward it. It surrounds a bacterium or small protist with its pseudopodia, forming a *food vacuole.* Enzymes move into the vacuole to digest the food, and the digested food passes out of the vacuole into the cytoplasm of the amoeba. To get rid of wastes, an amoeba reverses the process. A waste-filled vacuole is moved to the edge of the cell and is released. **Figure 12** shows an amoeba feeding.

Some amoebas are parasites. Certain species live in the human intestine and cause amebic dysentery, a painful condition that can involve bleeding ulcers.

WEIRD SCIENCE

People may have more in common with amoebas than they think. Researchers have discovered that very similar contractile proteins are involved in both amoeboid movement and the movement of animal muscles. The cytoplasm of an amoeba contains thick and thin filaments that are almost identical to the thick myosin filaments and the thin actin filaments found in striated muscle.

Protozoa with Shells Not all amoebalike protozoa look like amoebas. Some have an outer shell. *Radiolarians* (RAY dee oh LER ee uhnz) have shells made of silica that look like glass ornaments. This type of protozoan is shown in **Figure 13.** *Foraminiferans* (fuh RAM uh NIF uhr uhnz) have snail-like shells made of calcium carbonate.

Flagellates Flagellates (FLAJ uh LITS) are protozoa that use flagella to move. The flagella wave back and forth to propel the organism forward. Some flagellates live in water. Others are parasites that can cause disease.

The flagellate parasite, *Giardia lamblia,* lives in the digestive tracts of humans and other vertebrates. This parasite is shown in **Figure 14.** In an inactive form, *Giardia* (JEE ar DEE uh) can survive in water. Hikers or others who drink water infected with *Giardia* can get diarrhea and severe stomach cramps, but the disease is usually not fatal.

Some flagellates live in symbiosis with vertebrates or invertebrates. In *symbiosis,* one organism lives closely with another organism, and each organism helps the other survive. One symbiotic flagellate lives in the guts of termites and digests the cellulose in the wood that the termites eat. Without the protozoa, the termites could not completely digest the cellulose.

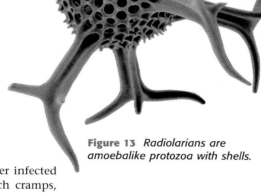

Figure 13 *Radiolarians are amoebalike protozoa with shells.*

Geology
C O N N E C T I O N

Foraminiferans have existed for over 600 million years. During this time, the shells of dead foraminiferans have been sinking to the bottom of the ocean. Millions of years ago, foraminiferan shells formed a thick layer of sediment of limestone and chalk deposits. The chalk deposits known as the White Cliffs of Dover in England were formed this way.

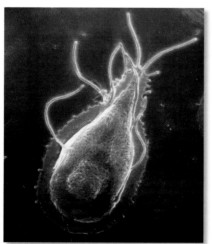

Figure 14 *The parasitic protist,* Giardia lamblia, *is a primitive cell. Can you see why it is a flagellate?*

53

IS THAT A FACT!

Deposits of foraminiferan shells on the sea floor are thousands of meters thick and cover millions of square kilometers. The sand of some beaches is also mostly from foraminiferan remains. There are nearly 50,000 foraminiferan shells in a gram of sand.

DISCUSSION

Protists and Pyramids Ask students:

What are the great pyramids of Egypt made of? (Accept all reasonable responses.)

Tell students that the pyramids are made, in part, of protist shells. Ask students how that could be. (Accept all reasonable responses.)

Tell them that tiny single-celled protists called foraminiferans make shells. When the foraminiferans die, the shells fall to the ocean floor. Over millions of years, the shells collect and mix with other minerals in the ocean water. Under the pressure of the ocean water, these components form the type of limestone that was used to make the great pyramids of Egypt.

CONNECT TO
EARTH SCIENCE

There are more than 40,000 recognized species of foraminiferans, but about 90 percent of those are extinct and known only from the fossil record. For millions of years, different species of foraminiferans flourished and then died out, leaving a record of their existence as distinctive layers in sea-floor deposits. In many parts of the world, some of those ancient sea-floor deposits have been uplifted and exposed as thick outcroppings of limestone, such as the White Cliffs of Dover. These fossil foraminiferans provide geologists with clues about the age of rock layers and how limestone deposits in different parts of the world are related. To review the geologic time scale with students, use Teaching Transparency 121.

Teaching Transparency 121
"The Geologic Time Scale"
LINK TO EARTH SCIENCE

GOING FURTHER

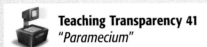

Anton van Leeuwenhoek (1632–1723) called the tiny, active organisms he viewed under his handmade microscopes *animalcules.* The ciliated protozoa undoubtedly are the most animal-like members of the kingdom Protista. Have students research a ciliate protozoan of their choice. Encourage students to make posters of their ciliate, complete with a large, detailed drawing of it and information about where it lives, what it eats, how it moves, and other interesting aspects of its biology. Ask permission to display students' posters in the school cafeteria or the library.

Answers to Self-Check

1. Cilia are used to move a ciliate through the water and to sweep food toward the organism.

2. Ciliates are classified as animal-like protists because they are consumers and they move.

Teaching Transparency 41 *"Paramecium"*

Ciliates Ciliates (SIL ee its) are the most complex protozoa. Ciliates have hundreds of tiny hairlike structures known as *cilia.* The cilia move a protozoan forward by beating back and forth. Cilia can beat up to 60 times a second! In some species, clumps of cilia form bristlelike structures used for movement. Cilia are also important for feeding. Ciliates use their cilia to sweep food through the water toward them. The best known ciliate is *Paramecium* (PAR uh MEE see uhm), shown in **Figure 15.**

Ciliates have two kinds of nuclei. A large nucleus called a *macronucleus* controls the functions of the cell. A smaller nucleus, the *micronucleus,* passes genetic material to another individual during sexual reproduction.

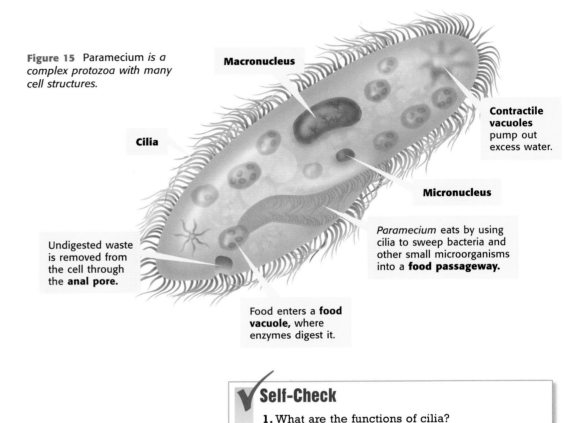

Figure 15 Paramecium *is a complex protozoa with many cell structures.*

Macronucleus

Cilia

Contractile vacuoles pump out excess water.

Micronucleus

Undigested waste is removed from the cell through the **anal pore.**

Paramecium eats by using cilia to sweep bacteria and other small microorganisms into a **food passageway.**

Food enters a **food vacuole,** where enzymes digest it.

 **Self-Check**

1. What are the functions of cilia?

2. Why are the ciliates classified as animal-like protists?

(See page 168 to check your answers.)

54

WEIRD SCIENCE

Not all ciliates feed on small prey. A barrel-shaped *Didinia* gobbles up protists that are much larger than itself. *Paramecium* is a common prey of *Didinia,* even though *Paramecium* is twice as large as *Didinia.* To accomplish this feat, *Didinia* essentially folds a paramecium in half and then engulfs it.

Spore-Forming Protists The spore-forming protozoa are all parasites that absorb nutrients from their hosts. They have no cilia or flagella, and they cannot move on their own. Spore-forming protozoa have complicated life cycles that usually involve two or more different hosts.

Plasmodium (plaz MOH dee uhm) *vivax* (VIE vaks) is the spore-forming protist that causes malaria. Malaria is a serious disease that is carried by mosquitoes in tropical areas. Although malaria can be treated with drugs, more than 2 million people die from malaria each year.

The Life Cycle of *Plasmodium vivax*

Plasmodium vivax is a parasite that has two different hosts, mosquitoes and humans. It needs both hosts to survive. Once a human is bitten by an infected mosquito, *Plasmodium* spores enter the liver. The spores multiply and change form.

They then enter red blood cells and multiply, causing the red blood cells to burst. If the malaria victim is bitten by another mosquito, the disease can then be carried to another human.

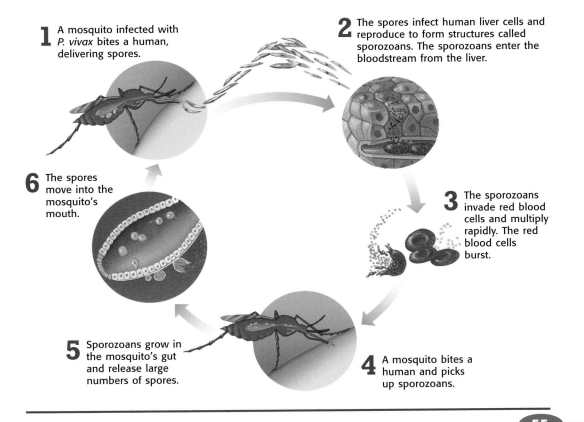

1 A mosquito infected with *P. vivax* bites a human, delivering spores.

2 The spores infect human liver cells and reproduce to form structures called sporozoans. The sporozoans enter the bloodstream from the liver.

3 The sporozoans invade red blood cells and multiply rapidly. The red blood cells burst.

4 A mosquito bites a human and picks up sporozoans.

5 Sporozoans grow in the mosquito's gut and release large numbers of spores.

6 The spores move into the mosquito's mouth.

MATH and MORE

Many protists reproduce asexually simply by dividing in two, a process called fission. Students are probably familiar with this form of division, which can result in a population growing exponentially over time; that is, 1 cell divides to produce 2 cells, which in turn divide to produce 4 cells, then 8, 16, 32, 64, and so on. *Plasmodium* and some other parasitic, spore-forming protozoa can divide asexually by a process known as schizogony, or multiple fission. For example, a single spore (technically called a sporozoite) of malaria-causing strain called *Plasmodium vivax* can produce 40,000 offspring.

Researchers have determined that when a *Plasmodium*-infected mosquito inserts its proboscis into a human blood vessel, it injects about a thousand spores. Have students calculate how many *Plasmodium vivax* spores could be present in a person's body after just one division by multiple fission.

(1,000 spores × 40,000 offspring = 40,000,000 spores)

Math Skills Worksheet "A Shortcut for Multiplying Large Numbers"

IS THAT A FACT!

After a person is bitten by a malaria-carrying mosquito, he or she can get malaria even if only one *Plasmodium* spore successfully reaches the liver and reproduces.

Teaching Transparency 42 "The Life Cycle of *Plasmodium vivax*"

Science Skills Worksheet "Organizing Your Research"

Homework

Writing **Research** Encourage students to research and report on malaria and the work currently being done on developing malaria vaccines.

Section 1 • Protists **55**

Quiz

1. What important characteristic do funguslike protists have in common with true fungi? (Sample answer: Both fungi and funguslike protists are consumers that obtain their food from dead organic matter or from the bodies of other organisms.)

2. Give three examples of protists that are producers in ocean food chains. (red, brown, and green algae; diatoms; dinoflagellates)

ALTERNATIVE ASSESSMENT

Writing | Ask students to create an illustrated book of the kingdom Protista, in which they devote a chapter to each of the major protist groups: funguslike protists, plantlike protists, and animal-like protists. Encourage students to use reference books and their textbook to create illustrations and write text for their protist books.

Reinforcement Worksheet "Protists on Parade"

Critical Thinking Worksheet "Protist Pop Culture"

Reproduction of Protists

Some protists reproduce asexually. In asexual reproduction, the offspring come from just one parent. Both animal-like amoebas and plantlike *Euglena* reproduce asexually by fission, as shown in **Figure 16.**

Some protists can also reproduce sexually. Sexual reproduction requires two parents. Animal-like *Paramecium* sometimes reproduces sexually by a process called conjugation. During conjugation, two *Paramecium* join together and exchange genetic material using their micronuclei. Then they divide to produce four organisms with new combinations of genetic material. Conjugation is shown in **Figure 17.**

Many protists reproduce both asexually and sexually. In some algae, asexual reproduction and sexual reproduction alternate from one generation to the next.

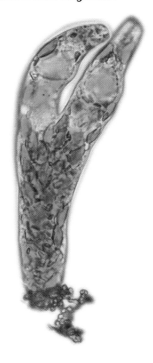

Figure 16 *During fission,* Euglena *divides lengthwise.*

Figure 17 *Conjugation in* Paramecium *is a type of sexual reproduction.*

internetconnect

SCiLINKS.
NSTA

TOPIC: Algae, Protozoa
GO TO: www.scilinks.org
*sci*LINKS NUMBER: HSTL255, HSTL260

SECTION REVIEW

1. Name the three main groups of protists, and give the characteristics of each.

2. What are three ways that flagella and cilia differ?

3. **Making Inferences** Killing mosquitoes is one method of controlling malaria. Using what you know about the organism that causes malaria, explain why this method works.

Answers to Section Review

1. **Funguslike protists** get their food from dead organic matter or from the body of another organism. **Plantlike protists** are producers that carry out photosynthesis. **Animal-like protists** are single-celled consumers.

2. Flagella are whiplike structures, and cilia are tiny hair-like structures. Organisms have a small number of flagella at specific locations on their body. Organisms with cilia have hundreds of cilia all over their outer surface. Flagella are used only in moving the organism. Cilia are also used for feeding.

3. Malaria requires two hosts—mosquitoes and humans. Because it must have both hosts to survive, removing the mosquitoes will keep the protist from infecting additional humans.

Fungi

Terms to Learn

fungus   spore
hyphae   mold
mycelium   lichen

What You'll Do

- ◆ Describe the characteristics of fungi.
- ◆ Distinguish between the four main groups of fungi.
- ◆ Describe how fungi can be helpful or harmful.
- ◆ Define *lichen*.

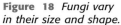

Have you ever heard someone say, "A fungus is among us"? This statement has more truth in it than you may realize. The mushrooms on pizza are a type of fungus (plural, *fungi*). The yeast used to make bread is a fungus. Fungi are also used to produce cheeses, antibiotics, and soy sauce. And if you've ever had athlete's foot, you can thank a fungus. Fungi are everywhere!

Characteristics of Fungi

Fungi are eukaryotic consumers, but they are so different from other organisms that they are placed in their own kingdom. As you can see in **Figure 18,** fungi come in a variety of shapes, sizes, and colors. But all fungi have similar ways of obtaining food and reproducing.

Figure 18 *Fungi vary in their size and shape.*

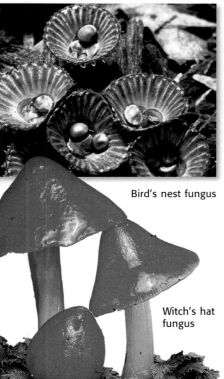

Bird's nest fungus

Witch's hat fungus

Straight coral fungus

Ascomycetus

Food for Fungi   Fungi are consumers, but they cannot eat or engulf food. Fungi must live on or near their food supply. Most fungi obtain nutrients by secreting digestive juices onto a food source, then absorbing the dissolved substances. Many fungi are *decomposers*. This means that they feed on dead plant or animal matter. Other fungi are parasites.

Some fungi live in symbiotic relationships with other organisms. For example, many types of fungi grow on the roots of plants. They release an acid that changes minerals in the soil into forms that plants can use. The fungi also protect the plant from some disease-causing organisms.

57

IS THAT A FACT!

There are about 350 species of yeast. The most economically important yeast is *Saccharomyces cerevisiae,* which has been used by humans in the production of bread, beer, and wine for thousands of years.

Directed Reading Worksheet Section 2

internetconnect

SCILINKS
NSTA

TOPIC: Fungi
GO TO: www.scilinks.org
***sci*LINKS NUMBER:** HSTL265

Focus

Fungi

In this section, students learn about fungi—eukaryotic consumers that obtain their food by absorbing nutrients from other organisms. Students are introduced to the four main groups of fungi: threadlike fungi, sac fungi, club fungi, and imperfect fungi. Finally, students learn about lichens, which are symbiotic associations between a fungus and an organism that can photosynthesize, usually an alga.

Bellringer

Ask students to answer the following questions in their ScienceLog:

What are mushrooms? What is the function of the cap of a button mushroom? (Accept all reasonable responses. The umbrella-shaped mushrooms that students may be most familiar with are club fungi. The above-ground portion of the mushroom is the spore-producing part of the organism.)

1) Motivate

DISCUSSION

Ask students to describe a world without fungi. (Accept all reasonable responses, but guide them to understand that without fungi there would be no leavened bread, no penicillin, no bleu cheese, no mushroom pizza. Tell them that fungi are decomposers and that huge amounts of dead organic matter might collect without fungi.)

Answers to Self-Check

1. Both fungi and funguslike protists are consumers. Both form spores.

2. Hyphae are chains of fungal cells that grow together to form a large mass called the mycelium.

READING 📖 STRATEGY

Prediction Guide Before students read the rest of this section, ask them if the following three statements are true or false. Students will learn the answers as they explore Section 2.

- Unlike protists, fungi can reproduce only asexually. (false)

- Morels and truffles are highly prized mushrooms. (true)

- Lichens are a combination of a fungus and an alga. (true)

DISCUSSION

Allergies and Molds Ask students if they or people they know are allergic to molds. As a class, encourage students to discuss reasons why mold allergies are so common, and the types of environments or substances that mold-sensitive individuals might want to avoid and why. (Sample answer: caves, damp basements, damp soil, leaf litter and other decaying organic matter, and so on; Molds flourish in such places and materials.)

✔ Self-Check

1. In what ways are fungi and funguslike protists alike?

2. How are hyphae and mycelia related?

(See page 168 to check your answer.)

Hidden from View All fungi are made of eukaryotic cells, which have nuclei. Some fungi are single-celled, but most fungi are multicellular. Multicellular fungi are made up of chains of cells called **hyphae** (HIE fee). Hyphae are fungal filaments that are similar to plant roots. These filaments are made of cells. But unlike plant root cells, the hyphae cells have openings in their cell walls that allow cytoplasm to move freely between the cells. The hyphae grow together to form a twisted mass called the **mycelium** (mie SEE lee uhm). The mycelium is the major part of the fungus, but it is often hidden from view underneath the ground. **Figure 19** shows the hyphae and mycelium of a fungus.

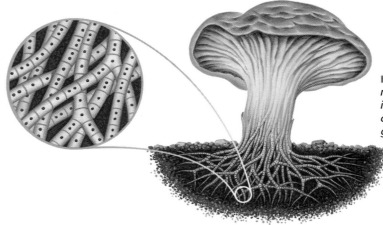

Figure 19 *The mycelium of a fungus is formed by hyphae and is often underground.*

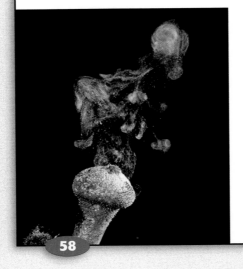

Making More Fungi Reproduction in fungi may be either asexual or sexual. Asexual reproduction occurs in two ways. In one type of asexual reproduction, the hyphae break apart and each new piece becomes a new individual. Asexual reproduction can also occur by the production of spores. **Spores** are small reproductive cells protected by a thick cell wall. Spores are light and easily spread by the wind. See for yourself in **Figure 20.** If the growing conditions where it lands are right, a spore will produce a new fungus.

Sexual reproduction occurs in fungi when special structures form to make sex cells. The sex cells join to produce sexual spores that grow into a new fungus.

Figure 20 *This puffball is releasing spores that can produce new fungi.*

IS THAT A FACT!

Yeasts are widely used in genetic research. Yeasts were the first eukaryotic organisms to have their DNA manipulated through the techniques of genetic engineering; today they are the organisms of choice for many types of experiments in molecular and cell biology. Experiments on yeasts such as *Saccharomyces cerevisiae,* the common bread yeast, have provided enormous insight into the functioning of eukaryotic cells.

Kinds of Fungi

Fungi are divided into four main groups: threadlike fungi, sac fungi, club fungi, and imperfect fungi. A fungus is classified into a particular group based on its shape and the way it reproduces.

Threadlike Fungi Have you ever seen fuzzy black mold growing on bread? **Molds** are shapeless fuzzy fungi, as shown in **Figure 21**. This particular mold belongs to a group of fungi called *threadlike fungi*. Most of the fungi in this group live in the soil and are usually decomposers, although some are parasites.

Threadlike fungi can reproduce asexually. Extensions of the hyphae grow into the air and form round spore cases at the tips called *sporangia* (spoh RAN jee uh). These sporangia are shown in **Figure 22**. When the sporangia break open, many tiny spores are released into the air.

Threadlike fungi can also reproduce sexually. Two hyphae from different individuals join and develop into specialized sporangia. These sporangia can survive periods of cold or drought. When conditions become more favorable, these specialized sporangia release spores that can grow into new fungi.

Figure 21 *Black bread mold is a soft, cottony mass that grows on bread and fruit.*

Figure 22 *Each of the round sporangia contains thousands of spores.*

Self-Check

What is the relationship between spores and sporangia? *(See page 168 to check your answers.)*

QuickLab

Moldy Bread

If you took a **slice of bread,** moistened it with a few drops of **water,** and then sealed it in a **plastic bag** for 1 week, what do you think would happen? Would the bread get moldy? Why or why not? Where would mold spores come from? How would these spores grow? Design an experiment to check your predictions. How many pieces of bread will you use? Will you treat them the same or differently? Why? If your teacher approves, try your experiment. Did it answer your questions? If not, what changes could you make to get the answers?

59

Multicultural CONNECTION

Centuries ago, Indonesians discovered that the bread mold *Rhizopus* can be used to produce a food called tempeh. Soybeans that have been stripped of their skins and boiled are inoculated with this mold and allowed to sit for 24 hours. A mycelium grows around the soybeans, holding them together in a mass and producing enzymes that increase the level of B vitamins in the mixture. Tempeh is very nutritious and is prepared daily in many Indonesian households, just as bread is in many other countries. Tempeh is usually served fried, baked, steamed, or roasted.

QuickLab

MATERIALS

FOR EACH GROUP:
• slice of bread
• water
• plastic bag

Answers to QuickLab

Answers will vary, but explanations and experimental designs should be logical. Students may suggest that the bread would become moldy. Mold spores would come from the air and grow using the nutrients in the bread and water. An experiment to test the hypothesis would have at least two groups—a plastic bag containing a dry piece of bread and a plastic bag containing a moistened slice of bread. Students may also suggest other variations, such as leaving one bag open and the other sealed.

Answer to Self-Check

Sporangia are round tips on hyphae extensions. Spores are found in the sporangia.

WEIRD SCIENCE

The mold *Pilobolus* grows on animal manure. The name comes from Greek words meaning "hat thrower." The name fits: *Pilobolus* produces little black sacs of spores on top of stalked structures that swell just beneath the sacs. As the spores mature, pressure builds up in the swollen parts of the stalks until spore-containing sacs are shot into the air, sometimes as much as 8 m! The sticky spore sacs adhere to grass and leaves. When animals eat the plants, *Pilobolus* spores pass unharmed through their digestive track and end up in their dung, where the spores germinate and begin the cycle again.

MAKING MODELS

Encourage students to research a species of sac or club fungus in an encyclopedia, in books on fungi, or on the Internet. Provide students with colored modeling clay, and have them create lifelike models of the fungi they have investigated. Make a class display of the fungus models. **Sheltered English**

MATH and **MORE**

Truffles are called the black diamonds of French cooking. Tell students that, in western France an unusually dry summer produced a smaller than usual harvest. Prices of first quality truffles rose from $535 per kilogram to $625 per kilogram.

Ask students the following question:

> What percentage increase in the price of truffles does this represent? ($625 − $535 = $90; $90 ÷ 535 = 0.168, or a 17% increase)

Tell students to imagine they are chefs and are required to make a new dish with a recipe calling for 1 lb of truffles. How many kilograms of this high-priced fungi will they need to buy, and what will it cost at the new price? (1 kg = 2.2 lb, so students will need 0.45 kg for this recipe. The cost would be $625 × 0.45 = $281.25.)

Figure 23 *Many people think truffles are delicious. Would you eat them?*

MATH BREAK

Multiplying Yeasts

Under ideal conditions, a yeast will produce a new cell by budding in about 30 minutes. Suppose a beaker contains 100 yeast cells. How many cells will it contain after 30 minutes? after 1 hour? after 2 hours? Make a graph to show the increase in size of the yeast population over a period of 5 hours.

Figure 25 *Dutch elm disease is a fungal disease that has killed thousands of elm trees in North America.*

60

Sac Fungi   *Sac fungi* form the largest group of fungi. Sac fungi include yeasts, powdery mildews, truffles, and morels. Truffles are shown in **Figure 23.**

Sexual reproduction in these fungi involves the formation of a sac called an *ascus*. These sacs give the sac fungi their name. Sexually produced spores develop within the ascus. During their life cycles, sac fungi usually reproduce both sexually and asexually.

Most sac fungi are multicellular, but *yeasts* are single-celled sac fungi. Yeasts reproduce asexually by *budding*. In budding, a new cell pinches off from an existing cell. A yeast is budding in **Figure 24.** Yeasts are the only fungi to reproduce by budding.

Figure 24 *Yeasts reproduce by budding. A round scar forms where a bud breaks off of a parent cell. How many times has the larger cell reproduced?*

Some sac fungi are very useful to humans. One example is yeasts, which are used in making bread. Yeasts use sugar as food and produce carbon dioxide gas and alcohol as waste products. Trapped bubbles of carbon dioxide cause the dough to rise and make bread light and fluffy. Other sac fungi are sources of antibiotics and vitamins. Truffles and morels are prized edible fungi.

Many sac fungi are parasites. They cause plant diseases, such as chestnut blight and Dutch elm disease, shown in **Figure 25.**

Answers to MATHBREAK

After 30 minutes, there will be 200 yeast cells. After 1 hour, there will be 400 cells, and after 2 hours, there will be 1,600 cells. The graph will show exponential growth to a point of 102,400 cells (5 hours).

IS THAT A FACT!

Some yeast produce an enzyme called invertase that is used by commercial candy makers to soften or liquefy the centers of chocolate candies after the coating has been applied.

Club Fungi The umbrella-shaped mushrooms are the most commonly known fungi. They belong to a group of fungi called *club fungi*. During sexual reproduction, special hyphae develop and produce clublike structures called *basidia* (buh SID ee uh), the Greek word for "clubs." Sexual spores develop inside the basidia.

What you think of as a mushroom is only the sexual spore-producing part of the organism. The mass of hyphae from which mushrooms are produced may grow 35 m across. Since mushrooms usually grow at the outer edges of the mass of hyphae, they often appear in circles, as shown in **Figure 26.**

The most familiar mushrooms are known as gill fungi because the basidia develop in the grooves, or *gills*, under the cap. Some varieties are grown commercially and sold in supermarkets, but not all gill fungi are edible. The white destroying angel is one type that is very poisonous. Simply a taste of this mushroom can be fatal. See if you can pick out the poisonous fungus in **Figure 27.**

Figure 26 A ring of mushrooms can appear overnight. In European folk legends, these were known as "fairy rings."

Figure 27 Many poisonous mushrooms look good to eat. The mushrooms on the left are edible, but the ones on the right are poisonous.

A Mushroom Omelet
A friend wants to make a mushroom omelet, but he has no mushrooms. He recently got a mushroom book that has pictures of all the poisonous and edible mushrooms. Using the book, he picks some mushrooms in the woods behind his house. He uses the wild mushrooms to make an omelet. Do you think he should eat the omelet? Why or why not?

Observe a Mushroom

1. Identify the stalk, cap, and gills on a **mushroom** that your teacher has provided.

2. Carefully twist or cut off the cap, and cut it open with a **knife.** Observe the gills with a **magnifying lens.** Look for spores.

3. Observe the other parts of the mushroom with the magnifying lens. The mycelium begins at the bottom of the stalk. Try to find individual hyphae.

4. Sketch the mushroom, and label the parts.

61

GUIDED PRACTICE

Writing Have students write the following words in their ScienceLog and match the type of fungus with its appropriate reproductive structure. Then have students use each correctly matched pair in a sentence:

threadlike fungi, sac fungi, club fungi ascus, basidium, zygospores

MISCONCEPTION ALERT

People often think that they can rid a lawn or a garden of mushrooms by pulling them out like weeds or chopping them up with a hoe whenever they appear. Because the largest part of a club fungus lives underground, this is wasted effort. In fact, chopping up mushrooms with a hoe may help to release and spread fungal spores, which could grow into additional mycelia.

QuickLab

MATERIALS
• mushroom
• plastic knife
• magnifying lens

Safety Caution: Remind students to review all safety cautions and icons before beginning this lab activity. Tell students not to eat the mushrooms.

Answers to QuickLab

Students should be able to find the stalk, cap, and gills. Spores and hyphae may or may not be visible.

Answer to APPLY

He should not eat the omelet. Even though he used a guide book to select the mushrooms, many poisonous mushrooms look very similar to edible mushrooms.

WEIRD SCIENCE

Mushrooms consist mostly of water (about 90 percent), and most edible species are low in nutritional value. An exception is the shiitake mushroom, which has been grown for centuries in Japan and China. In ancient China, people believed that eating shiitake mushrooms promoted good health. They were right—shiitake mushrooms are rich in iron, phosphorus, and calcium and in vitamins B, D, and C. They also have twice the protein content of most other commercially grown mushrooms.

3) Extend

CROSS-DISCIPLINARY FOCUS

Writing **Literature** Beatrix Potter (1866–1943) is best known for authoring and illustrating such famous stories as *The Tale of Peter Rabbit* and *The Tale of Mr. Toad.* She lived and worked in England and had a scholarly interest in fungi. Some of her papers were unfairly rejected, and she detailed this in her diaries. She was a shy person, easily dismissed by the scholars at the Royal Society and the Royal Botanical Gardens. After all, what could the inventor of Peter Rabbit know about fungi? She is now widely respected as a mycologist (a scientist who studies fungi) and is recognized for contributing many valuable papers on the subject. Her detailed drawings of more than 270 examples of fungi are displayed in a library in Ambleside, England. Encourage interested students to research and present a report on the life of Beatrix Potter.

PORTFOLIO

REAL-WORLD CONNECTION

Cyclosporin is a chemical compound that was discovered in an imperfect fungus in the 1980s. Cyclosporin suppresses immune reactions that lead to rejection of transplanted organs. This fungal drug has led to much greater success in organ transplants.

Figure 28 *Bracket fungi look like shelves on trees. Spores are found on the underside of the bracket.*

Mushrooms are not the only club fungi. Bracket fungi, puffballs, smuts, and rusts are also in this group of fungi. Bracket fungi grow outward from wood, forming small shelves or brackets, as shown in **Figure 28.** Smuts and rusts are common plant parasites. They often attack crops such as corn and wheat. This can be seen in **Figure 29.**

Figure 29 *This corn crop is infected with a club fungus called a smut.*

BRAIN FOOD

Did you know that stone-washed jeans aren't really washed with stones? They get their faded look from a fungus! Jeans are soaked in a solution containing the fungus *Trichoderma.* This fungus produces enzymes that partially digest the cotton fibers to give jeans a stone-washed appearance.

Imperfect Fungi The *imperfect fungi* group includes all the species of fungi that do not quite fit in the other groups. These fungi do not reproduce sexually. Most are parasites that cause diseases in plants and animals. One common human disease caused by these fungi is athlete's foot, a skin disease. Another fungus from this group produces a poison called *aflatoxin,* which can cause cancer.

Some imperfect fungi are useful. *Penicillium,* shown in **Figure 30,** is the source of the antibiotic penicillin. Other imperfect fungi are also used to produce medicines. Some imperfect fungi are used to produce cheeses, soy sauce, and the citric acid used in cola drinks.

Figure 30 *The fungus* Penicillium *produces a substance that kills certain bacteria.*

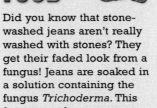

WEIRD SCIENCE

Near Seattle, Washington, a fungus was found that covered an area of 4 km² (2.5 mi²) and weighed nearly 1,000 tons. When scientists took tissue samples of this honey mushroom fungus, they found it to be one organism! The fungus is estimated to be up to 1,000 years old.

IS THAT A FACT!

If you are allergic to penicillin, you will probably be allergic to Camembert and Roquefort cheeses too, because these cheeses get their flavor from *Penicillium* molds.

Lichens

A **lichen** is a combination of a fungus and an alga that grow intertwined. The alga actually lives inside the protective walls of the fungus. The resulting organism is different from either of the two organisms growing alone. The merging of the two organisms to form a lichen is so complete that scientists give lichens their own scientific names. **Figure 31** shows examples of lichens.

Unlike fungi, lichens are producers. The algae in the lichens produce food through photosynthesis. Unlike algae, lichens can withstand drying out because of the protective walls of the fungus. Lichens are found in almost every type of terrestrial environment. They can even grow in extreme environments like dry deserts and the Arctic.

Lichens need only air, light, and minerals to grow. This is why lichens can grow on rocks. They produce acids that break down the rock and cause cracks. Bits of rock and dead lichens fill the cracks, making soil that other organisms can grow on.

Lichens absorb water and minerals from the air. As a result, they are easily affected by air pollution. Thus, the presence or absence of lichens is a good measure of air quality in an area.

Jewel lichen

Fruticose lichen

British soldier lichen

Figure 31 *These are some of the many types of lichens.*

SECTION REVIEW

1. How are fungi able to withstand periods of cold or drought?

2. Why are fungi such an important part of the natural world?

3. What are the four main groups of fungi? Give a characteristic of each.

4. **Making Inferences** Why are lichens an example of symbiosis?

internet**connect**

SCI**LINKS**
NSTA

TOPIC: Fungi, Lichens
GO TO: www.scilinks.org
*sci***LINKS NUMBER:** HSTL265, HSTL270

63

▼ **Answers to Section Review**

1. Fungi form spores that are able to withstand cold or drought.

2. Fungi are decomposers. They break down dead organic material, providing nutrients to the soil. Some fungi live in symbiotic relationships with other organisms. Some help plants grow by providing nutrients.

3. **Threadlike fungi** have hyphae that end in sporangia. **Sac fungi** have spores in a sac called an ascus. **Club fungi** produce club-like structures called basidia. **Imperfect fungi** do not reproduce sexually.

4. Lichens are a combination of a fungus and an alga. Each organism benefits from the other, and neither of the organisms can exist without the other, so it is a symbiotic relationship.

4) Close

Quiz

1. Why is it a good idea never to eat wild mushrooms that have not been identified positively by an expert? (Many poisonous mushrooms look harmless and closely resemble mushrooms that are safe to eat. It takes an expert to tell the difference.)

2. From what kind of a fungus is the antibiotic penicillin derived? (an imperfect fungus, *Penicillium*)

3. Where are lichens found? (They are found on bare rocks, soil, tree trunks, and mountain peaks in environments ranging from dry deserts to cold polar regions.)

ALTERNATIVE ASSESSMENT

Writing In their ScienceLog, have students write a narrative in which they describe several types of fungi they might expect to encounter on a walk through a temperate forest. Students should identify the group to which each fungus belongs and its role in the forest ecosystem. Encourage students to make illustrations or photomontages to accompany their narrative.

PORTFOLIO

Reinforcement Worksheet
"An Ode to a Fungus"

internet**connect**

SCI**LINKS**
NSTA

TOPIC: Lichens
GO TO: www.scilinks.org
*sci***LINKS NUMBER:** HSTL270

There's a Fungus Among Us!
Teacher's Notes

Time Required
One 45-minute class period

Lab Ratings

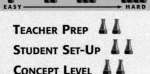

EASY ———————————→ HARD

TEACHER PREP 🧪🧪
STUDENT SET-UP 🧪🧪
CONCEPT LEVEL 🧪🧪
CLEAN UP 🧪🧪

Safety Caution
Before beginning this lab, check for any mushroom allergies among students. Caution students not to eat any of the mushrooms.

Preparation Notes
Do not use mushrooms gathered from the wild. Suitable mushrooms can be found in the produce section of a grocery store. Have protective gloves available for students who wish to wear them. This lab works well with groups of 2 to 4 students per mushroom. If you do not have an incubator, place the Petri dishes in a warm place, out of drafts and direct sunlight.

Skill Builder Lab

There's a Fungus Among Us!

Fungi share many characteristics with plants. For example, most fungi live on land. But fungi have several unique features that suggest that they are not closely related to any other kingdom of organisms. In this activity, you will observe some of the unique structures of a mushroom, a member of the kingdom Fungi.

MATERIALS

- mushroom
- tweezers
- 2 sheets of white paper
- masking tape
- Petri dish with fruit-juice agar plate
- incubator
- microscope or magnifying lens
- transparent tape

Procedure

1 Put on your safety goggles and get a mushroom from your teacher. Carefully pull the cap of the mushroom from the stem. Using tweezers, remove a gill from the underside of the cap. Place the gill on a sheet of white paper. Place the mushroom cap gill-side down on the other sheet of paper. Tape the cap in place with masking tape. Place the paper aside for at least 24 hours.

2 Use tweezers to take several 1 cm pieces from the stem, and place them on your agar plate. Record the appearance of the plate in your ScienceLog. Cover the Petri dish and incubate overnight.

3 Use tweezers to gently pull the remaining mushroom stem apart lengthwise. The individual fibers or strings you see are the hyphae that form the structure of the fungus. Place a thin strand on the same piece of paper with the gill you removed from the cap.

4 Observe the gill and the stem hyphae with a magnifying lens or microscope.

5 After at least 24 hours, record in your ScienceLog any changes that occurred on the agar. Carefully lift the mushroom cap from the paper. Place a piece of transparent tape over the print left behind on the paper. Record your observations in your ScienceLog.

64

Analysis

6 Describe the structures you saw on the gill and hyphae.

7 What is the print on the white paper?

8 Describe the structure at the bottom edge of the mushroom gill. Explain how this structure is connected to the print.

9 Explain how the changes that occurred in your Petri dish are related to methods of fungal reproduction.

65

Skill Builder Lab

Answers

6. Possible answer: Gills look like ridges and slits. Hyphae are stringlike.

7. The print on the white paper is a spore print produced by basidiospores being released from the gills.

8. The structures at the bottom edge of the mushroom gill are basidia, which contain the nuclei that will develop into the basidiospores.

9. After 24 hours, students may see mycelia growth on the agar plate. (It may take a little longer.) The mycelia constitute the main body of the mushroom and develop from the germinating basidiospores.

Datasheets for LabBook

Jason Marsh
Montevideo High and Country School
Montevideo, Minnesota

Chapter Highlights

Chapter Highlights

VOCABULARY DEFINITIONS

SECTION 1

protist an organism that belongs to the kingdom Protista

funguslike protist describes a protist that obtains its food from dead organic matter or from the body of another organism

parasite an organism that feeds on another living creature, usually without killing it

host an organism on which a parasite lives

algae protists that convert the sun's energy into food through photosynthesis

phytoplankton a microscopic photosynthetic organism that floats near the surface of the ocean

protozoa animal-like protists that are single-celled consumers

SECTION 1

Vocabulary

protist (p. 46)
funguslike protist (p. 47)
parasite (p. 48)
host (p. 48)
algae (p. 48)
phytoplankton (p. 49)
protozoa (p. 52)

Section Notes

• The protists are a diverse group of single-celled and multicellular organisms. They are grouped in their own kingdom because they differ from other organisms in many ways.

• Funguslike protists are consumers that obtain their food from dead organic matter or from the body of another organism.

• Slime molds and water molds are two groups of funguslike protists.

• Plantlike protists are producers that are also known as algae. Most are aquatic.

• Among the plantlike protists are red algae, brown algae, green algae, diatoms, dinoflagellates, and euglenoids.

• The animal-like protists are single-celled consumers also known as protozoa. Most can move.

• The protozoa include amoebalike protists, flagellates, ciliates, and spore-forming protists.

• Some protists reproduce sexually, some asexually, and some both sexually and asexually.

☑ Skills Check

Math Concepts

MICROBE MULTIPLICATION Suppose an amoeba can reproduce by fission once every 30 minutes. If you start with 50 amoebas, after 30 minutes you will have twice as many because each amoeba has divided.

$$2 \times 50 = 100 \text{ amoebas}$$

After 1 hour (30 minutes later) the number of amoebas will double again.

$$2 \times 100 = 200 \text{ amoebas}$$

Visual Understanding

PROTIST STRUCTURE Look at the illustration of *Paramecium* on page 54. Carefully read the labels, and look at each part described. Now do the same thing with the illustration of the euglenoid on page 51. Notice what the two cells have in common and how they differ.

66

Lab and Activity Highlights

There's a Fungus Among Us!   **PG 64**

Datasheets for LabBook
(blackline masters for these labs)

Vocabulary

fungus *(p. 57)*

hyphae *(p. 58)*

mycelium *(p. 58)*

spore *(p. 58)*

mold *(p. 59)*

lichen *(p. 63)*

Section Notes

- Fungi are consumers. They can be decomposers or parasites, or they can live in symbiotic relationships with other organisms.

- Most fungi are made up of chains of cells called hyphae. Many hyphae join together to form a mycelium.

- The four main groups of fungi are threadlike fungi, sac fungi, club fungi, and imperfect fungi.

- Threadlike fungi are primarily decomposers that form sporangia to hold spores.

- Molds are shapeless, fuzzy fungi.

- During sexual reproduction, sac fungi form little sacs in which sexual spores develop.

- Club fungi form structures called basidia during sexual reproduction.

- The imperfect fungi include all the species that do not quite fit anywhere else. Most are parasites that reproduce only by asexual reproduction.

- A lichen is a combination of a specific fungus and a specific alga that is different from either organism growing alone.

VOCABULARY DEFINITIONS, *continued*

SECTION 2

fungus an organism in the kingdom Fungi

hyphae chains of cells that make up multicellular fungi

mycelium a twisted mass of fungal hyphae that have grown together

spore a small reproductive cell protected by a thick wall

mold a shapeless, fuzzy fungus

lichen combination of a fungus and an alga that grow intertwined and exist in a symbiotic relationship

Vocabulary Review Worksheet

Blackline masters of these Chapter Highlights can be found in the **Study Guide.**

---

internet**connect**

 GO TO: go.hrw.com

Visit the **HRW** Web site for a variety of learning tools related to this chapter. Just type in the keyword:

KEYWORD: HSTPRO

SC/LINKS ᴼᴹ
N S T A

GO TO: www.scilinks.org

Visit the **National Science Teachers Association** on-line Web site for Internet resources related to this chapter. Just type in the *sci*LINKS number for more information about the topic:

TOPIC: Algae	*sci***LINKS NUMBER:** HSTL255
TOPIC: Protozoa	*sci***LINKS NUMBER:** HSTL260
TOPIC: Fungi	*sci***LINKS NUMBER:** HSTL265
TOPIC: Lichens	*sci***LINKS NUMBER:** HSTL270

67

Lab and Activity Highlights

LabBank

 Labs You Can Eat, Knot Your Average Yeast Lab

Long-Term Projects & Research Ideas, Algae for All!

Whiz-Bang Demonstrations, Unleash the Yeast!

Chapter Review
Answers

USING VOCABULARY

1. algae
2. conjugation
3. an ascus
4. Decomposers
5. protozoa
6. host

UNDERSTANDING CONCEPTS

Multiple Choice

7. d
8. a
9. c
10. b
11. b
12. d
13. b

Short Answer

14. Helpful fungi are decompsers that recycle nutrients. Some provide food for humans and are also used to make some medicines. Harmful fungi cause diseases in humans and damage crops.
15. *Paramecium* uses cilia to move and to get food.
16. A red tide is formed when red dinoflagellates multiply rapidly and form reddish brown water.
17. Both slime molds and amoebas are protists, and both feed by engulfing food.

Chapter Review

USING VOCABULARY

To complete the following sentences, choose the correct term from each pair of terms listed below:

1. Protists that get energy from photosynthesis are __?__. (*algae* or *amoebas*)

2. *Paramecium* reproduces sexually by __?__. (*budding* or *conjugation*)

3. The structure containing spores in a sac fungi is called __?__. (*an ascus* or *a basidium*)

4. __?__ live on dead organic matter. (*Parasites* or *Decomposers*)

5. Animal-like protists are also called __?__. (*protozoa* or *algae*)

6. A parasite gets its nutrients from its __?__. (*host* or *spores*)

UNDERSTANDING CONCEPTS

Multiple Choice

7. Plantlike protists include
   a. euglenoids and ciliates.
   b. lichens and flagellates.
   c. spore-forming protists and smuts.
   d. dinoflagellates and diatoms.

8. Funguslike protists
   a. are consumers or decomposers.
   b. are made of chains of cells called hyphae.
   c. are divided into four major groups.
   d. are always parasites.

9. A euglenoid has
   a. a micronucleus.
   b. pseudopodia.
   c. two flagella.
   d. cilia.

10. Fungi
    a. are producers.
    b. cannot eat or engulf food.
    c. are found only in the soil.
    d. are primarily single-celled.

11. A lichen
    a. is a parasite.
    b. is made up of an alga and a fungus that live intertwined together.
    c. can live only where there is plenty of water.
    d. is a consumer.

12. Animal-like protists
    a. are also known as protozoa.
    b. include amoebas and *Paramecium*.
    c. may be either free living or parasitic.
    d. All of the above

13. A contractile vacuole
    a. is a food passageway.
    b. pumps out excess water.
    c. is the location of food digestion.
    d. can be found in any animal-like protist.

Short Answer

14. How are fungi helpful to humans? How are they harmful?

15. What is the function of cilia in *Paramecium*?

16. What is a red tide?

17. How are slime molds and amoebas similar?

Concept Mapping

18. Use the following terms to create a concept map: fungi, ascus, club fungi, basidia, bread mold, yeast, threadlike fungi, mushrooms.

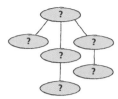

CRITICAL THINKING AND PROBLEM SOLVING

Write one or two sentences to answer the following questions:

19. What might happen if all the protists on Earth died?

20. Water can pass easily through a cell membrane, but too much water can cause a cell to burst. Single-celled amoebas and *Paramecium* have thin cell membranes, and they also live in water. Why don't these organisms burst?

21. You discover some mushrooms in your backyard one morning, and you pull them out of the ground. But the next day you find even more mushrooms. They seem to be growing in a line. Where did the new mushrooms come from?

MATH IN SCIENCE

22. You and your classmates are studying the growth of *Euglena*. Using a microscope, you count the number of organisms in a small container each day. Your teacher says you can expect the number of *Euglena* to double in 1.5 days. If you started out with five *Euglena*, how many *Euglena* do you think you will see on the third day?

INTERPRETING GRAPHICS

Look at the pictures of fungi below, and answer the following questions:

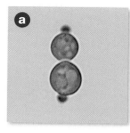
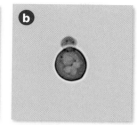

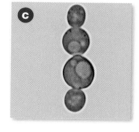
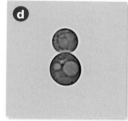

23. What kind of fungus is shown here?

24. What cellular process is shown in these pictures?

25. Which picture was taken first? last? Arrange the pictures in order.

26. Which is the original parent cell? How do you know?

Reading Check-up

Take a minute to review your answers to the Pre-Reading Questions found at the bottom of page 44. Have your answers changed? If necessary, revise your answers based on what you have learned since you began this chapter.

Concept Mapping

18. An answer to this exercise can be found at the front of this book.

CRITICAL THINKING AND PROBLEM SOLVING

19. Sample answer: If all the protists on Earth died, many larger organisms that depend on protists for food might also die. The oxygen supply on Earth might be severely limited if there were no plantlike protists.

20. Amoebas and *Paramecium* have contractile vacuoles that pump excess water outside of the cell.

21. The new mushrooms grew from mycelia that are hidden below the ground.

MATH IN SCIENCE

22. 20 *Euglena*

INTERPRETING GRAPHICS

23. yeast (sac fungus)
24. budding
25. b was first; c was last; b, d, a, c
26. The large cell in picture b is the parent cell because it is the first to bud.

 Concept Mapping Transparency 11

Blackline masters of this Chapter Review can be found in the **Study Guide.**

69

Background

The research into fungi-based dressing is being done nearly exclusively by the British Textile Technology Group. More-detailed descriptions of the research can be found on the Internet.

Teaching Strategy

Have the students research a folk remedy and create a poster depicting the ailment and the treatment.

Science, Technology, and Society

Moldy Bandages

When you think of the word *fungus,* you probably think of moldy leftovers in the refrigerator or mushrooms growing at the base of a tree. You may even think of athlete's foot or some other ailment caused by a fungus. Someday you may also think of bandages when you think of fungi. At least that is the hope of Paul Hamlyn and his colleagues at the British Textile Technology Group (BTTG).

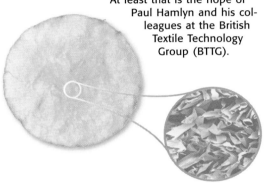

▲ *Would you believe that fungi like this may someday be used in surgical bandages?*

Fungi Versus Infection

The scientists at BTTG, along with scientists at the Welsh School of Pharmacy, have discovered that the cell walls of fungi contain polymers that promote the growth of the human cells responsible for rebuilding tissue around a wound. These human healing cells are called fibroblasts, and studies show that the fungal polymers attract and help bind the fibroblasts at a wound. Researchers believe that the polymers react with oxygen to produce hydrogen peroxide. The hydrogen peroxide activates white blood cells and promotes the growth of the fibroblasts. The white blood cells help fight infection around the wound.

From Crab Shells to Wound Healing

Until recently, the only sources of these polymers were crab and prawn shells. The quality and quantity of polymers found in the shells of crabs and prawns vary with weather conditions and with the seasons. Hamlyn's studies have shown that fungi can provide a more consistent product. For even more consistent results, scientists are able to grow fungi in the laboratory in a liquid growth medium.

Where Do We Go from Here?

Although his work is still in the research stage, Hamlyn is working toward a commercial application of the wound-healing fungi. He is researching the possibility of manufacturing two types of bandages from fungi. The first would be made by freeze-drying pieces of the fungi to create an absorbent dressing. This could be used for patients with deep wounds. The second type would be used to help patients with bed sores or diabetic skin ulcers. This bandage would involve a wet dressing of fungi that could be placed over sores or abrasions on the surface of the skin. The fungi would help accelerate the healing process.

Fungi Find the Cure

▶ Fungal products have long been used in the field of medicine. One modern use is cyclosporin. It is a drug used to help prevent the rejection of transplanted organs in humans. Other uses for fungi include a popular folk remedy—placing moldy bread on a wound to promote healing. Research these or other types of medicines that are made from fungi to discover how they work.

Answers to Fungi Find the Cure

Accept all reasonable answers. There are many books that provide lists of such home remedies. One such series is *Chicken Soup and Other Folk Remedies,* by Joan and Lydia Wilen.

ACROSS ᴛʜᴇ SCIENCES

L I F E   S C I E N C E   •   E A R T H   S C I E N C E

It's Alive!

The Maya of Mexico and their descendants believe that Cueva de Villa Luz (the Cave of the Lighted House) is inhabited by powerful spirits. For centuries, they have walked past slimy globs that drip from the cave's ceiling without even thinking about them. When scientists decided to analyze these slime balls, they discovered that the formations are home to billions of microscopic organisms! They nicknamed these colonies "snot-tites" because they resemble mucus.

Life in Battery Acid

As people climb down into the pitch-dark passages of Cueva de Villa Luz, they are greeted by the stench of rotten eggs, an odor rarely found in caves and quickly recognized as potentially deadly hydrogen sulfide gas. This foul and dangerous gas is emitted by the snot-tites! Because it is not safe to remain inside the caverns if there is too much hydrogen sulfide, explorers must constantly monitor the level of the gas in the air.

A closer inspection of snot-tite drippings reveals that they contain sulfuric acid. When sulfuric acid dripped on some of the explorers' clothes, the clothes dissolved right off of their backs!

How do the organisms live in such harsh conditions? It turns out that snot-tites can actually get energy from sulfur, which is toxic to most organisms. And since snot-tites do not have to rely on photosynthesis for energy, they can live in absolute darkness.

A Unique Ecosystem

During dry seasons, the Mayan people feast on tiny fish called mollies that are abundant in the milky-white streams that flow through Cueva de Villa Luz. It's very rare to find so many fish in cave streams. Why are there fish in this cave?

Scientists have discovered that the snot-tites are part of a complex underground ecosystem, possibly unlike any other on Earth. When snot-tites use the sulfur found in the cave, they produce a nutritionally rich waste product. This waste drips into the streams below, where hungry fish can eat it.

Life on Mars?

Many scientists argue that conditions on Mars are too harsh to support any form of life. However, Martian rocks have a large amount of sulfur and scientists know that Mars has caves. Snot-tites have proven that life can exist in these harsh conditions. Someday a space traveler might find a similar organism on Mars!

Going to Extremes

▶ Snot-tites have adapted to extreme environmental conditions that would kill other organisms. Such organisms are called extremophiles. Investigate to find out about other extremophiles. Why are scientists so interested in extremophiles?

◀ *Several billion sulfur-eating microbes can live in a single cubic centimeter of these slimy, gooey snot-tites!*

71

Snot-tites, which are a mixture of fungi and bacteria, prove that life can exist in the most brutal environments. Scientists study extremophiles because they believe that the adaptations of organisms to tolerate harsh conditions may have practical applications. Most extremophiles have special enzymes or biological catalysts that allow them to survive in such extreme conditions. For example, enzymes allow snot-tites to convert sulfur, which kills most organisms, into cellular energy.

Heat-loving microorganisms, called thermophiles, can thrive in boiling temperatures of 212°F! Scientists are still trying to figure out how the enzymes of these organisms stay in one piece. Encourage students to investigate the thermophiles that live in the hot springs of Yellowstone National Park.

Answer to Going to Extremes

The enzymes of extremophiles are used industrially to produce sweeteners and to diagnose some diseases. Unfortunately, most enzymes fall apart and stop working when they are exposed to heat or harsh chemicals, such as acid. Scientists must take great steps to ensure that enzymes stay intact. Enzymes from extremophiles, nicknamed "extremozymes," may eliminate these difficult or costly steps.

Chapter Organizer

CHAPTER ORGANIZATION	TIME MINUTES	OBJECTIVES	LABS, INVESTIGATIONS, AND DEMONSTRATIONS
Chapter Opener pp. 72–73	45	National Standards: UCP 1, 2, 5, SPSP 5, LS 1a, 4c	**Start-Up Activity,** Observing Plant Growth, p. 73
Section 1 **What Makes a Plant a Plant?**	90	▶ Identify the characteristics that all plants share. ▶ Discuss the origin of plants. ▶ Explain how the four main groups of plants differ. UCP 1, 2, 5, SAI 1, 2, ST 2, HNS 2, LS 1a, 1d, 2b, 4c, 5a	**Demonstration,** Water Travel in Plants, p. 74 in ATE **Interactive Explorations CD-ROM,** Shut Your Trap! A **Worksheet** is also available in the **Interactive Explorations Teacher's Guide.**
Section 2 **Seedless Plants**	90	▶ Describe the features of mosses and liverworts. ▶ Describe the features of ferns, horsetails, and club mosses. ▶ Explain how plants without seeds are important to humans and to the environment. UCP 1–3, 5, SAI 1, LS 1a, 1c, 2b, 5c	**QuickLab,** Moss Mass, p. 79
Section 3 **Plants with Seeds**	90	▶ Compare a seed with a spore. ▶ Describe the features of gymnosperms. ▶ Describe the features of angiosperms. ▶ List the economic and environmental importance of gymnosperms and angiosperms. UCP 1, 2, 5, SAI 1, SPSP 4, 5, LS 1a, 1d, 2b–2d, 4b–4d; Labs UCP 2, 5, SAI 1, HNS 1, LS 1a, 5b	**Skill Builder,** Travelin' Seeds, p. 137 **Datasheets for LabBook,** Travelin' Seeds, Datasheet 26 **EcoLabs & Field Activities,** The Case of the Ravenous Radish, EcoLab 3
Section 4 **The Structures of Seed Plants**	90	▶ Describe the functions of roots. ▶ Describe the functions of stems. ▶ Explain how the structure of leaves is related to their function. ▶ Identify the parts of a flower and their functions. UCP 1, 2, 5, SAI 1, LS 1a, 1d, 2b, 3d, 4c, 5b; Labs UCP 2, 5, SAI 1, HNS 1, LS 1a, 2b, 5b	**Demonstration,** Seed Transport, p. 88 in ATE **Skill Builder,** Leaf Me Alone! p. 000 **Datasheets for LabBook,** Leaf Me Alone! Datasheet 25 **Making Models,** Build a Flower, p. 000 **Datasheets for LabBook,** Build a Flower, Datasheet 27 **Whiz-Bang Demonstrations,** Inner Life of a Leaf, Demo 8 **Long-Term Projects & Research Ideas,** Project 12

See page **T23** *for a complete correlation of this book with the*

NATIONAL SCIENCE EDUCATION STANDARDS.

TECHNOLOGY RESOURCES

 Guided Reading Audio CD English or Spanish, Chapter 4

 One-Stop Planner CD-ROM with Test Generator

 CNN. **Eye on the Environment,** Prairie Restoration, Segment 16

 Interactive Explorations CD-ROM CD 1, Exploration 2, Shut Your Trap!

CLASSROOM WORKSHEETS, TRANSPARENCIES, AND RESOURCES	SCIENCE INTEGRATION AND CONNECTIONS	REVIEW AND ASSESSMENT
Science Puzzlers, Twisters & Teasers **Directed Reading Worksheet**		
Directed Reading Worksheet, Section 1 **Transparency 43,** Plant Life Cycle **Transparency 44,** The Main Groups of Living Plants	**Multicultural Connection,** p. 76 in ATE **Math and More,** Percentages, p. 76 in ATE **MathBreak,** Practice with Percents, p. 77 **Careers:** Ethnobotanist—Paul Cox, p. 103	**Self-Check,** p. 76 **Review,** p. 77 **Quiz,** p. 77 in ATE **Alternative Assessment,** p. 77 in ATE
Directed Reading Worksheet, Section 2	**Cross-Disciplinary Focus,** p. 78 in ATE **Real-World Connection,** p. 79 in ATE	**Homework,** p. 80 in ATE **Review,** p. 81 **Quiz,** p. 81 in ATE **Alternative Assessment,** p. 81 in ATE
Directed Reading Worksheet, Section 3 **Science Skills Worksheet,** Science Writing **Transparency 45,** Two Classes of Angiosperms **Reinforcement Worksheet,** Classifying Plants **Reinforcement Worksheet,** Drawing Dicots	**Environment Connection,** p. 83 **Apply,** p. 83 **Connect to Earth Science,** p. 84 in ATE **Cross-Disciplinary Focus,** p. 86 in ATE **Science, Technology, and Society:** Supersquash or Frankenfruit? p. 102	**Homework,** p. 85 in ATE **Review,** p. 87 **Quiz,** p. 87 in ATE **Alternative Assessment,** p. 87 in ATE
Directed Reading Worksheet, Section 4 **Transparency 46,** Root Structure **Science Skills Worksheet,** Taking Notes **Transparency 47,** Stem Structure **Transparency 48,** Leaf Structure **Transparency 235,** Photosynthesis **Transparency 48,** Flower Structure **Critical Thinking Worksheet,** The Voodoo Lily	**Cross-Disciplinary Focus,** p. 90 in ATE **Connect to Physical Science,** p. 93 in ATE	**Review,** p. 91 **Self-Check,** p. 92 **Homework,** p. 92 in ATE **Review,** p. 95 **Quiz,** p. 95 in ATE **Alternative Assessment,** p. 95 in ATE

END-OF-CHAPTER REVIEW AND ASSESSMENT

Chapter Review in Study Guide
Vocabulary and Notes in Study Guide
Chapter Tests with Performance-Based Assessment, Chapter 4 Test
Chapter Tests with Performance-Based Assessment, Performance-Based Assessment 4
Concept Mapping Transparency 12

 Holt, Rinehart and Winston On-line Resources

go.hrw.com

For worksheets and other teaching aids related to this chapter, visit the HRW Web site and type in the keyword: **HSTPL1**

 National Science Teachers Association

www.scilinks.org

Encourage students to use the *sci*LINKS numbers listed in the internet connect boxes to access information and resources on the **NSTA** Web site.

Chapter Resources & Worksheets

Visual Resources

TEACHING TRANSPARENCIES

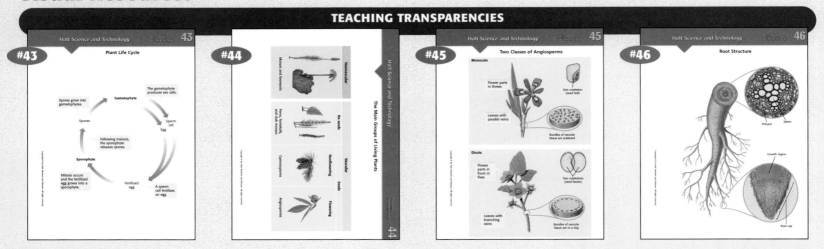

#43 Plant Life Cycle

#44 The Main Groups of Living Plants

#45 Two Classes of Angiosperms

#46 Root Structure

TEACHING TRANSPARENCIES

CONCEPT MAPPING TRANSPARENCY

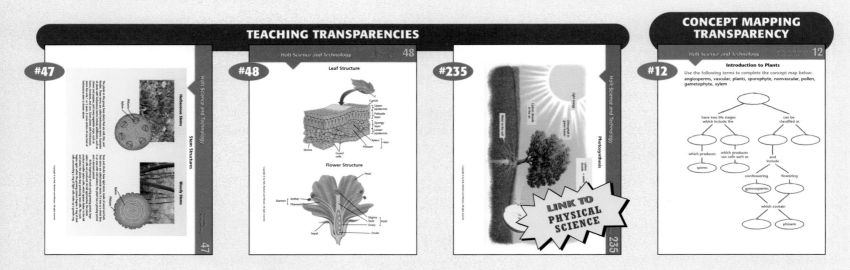

#47 Stem Structures

#48 Leaf Structure / Flower Structure

#235 Photosynthesis — LINK TO PHYSICAL SCIENCE

#12 Introduction to Plants

Meeting Individual Needs

DIRECTED READING

REINFORCEMENT & VOCABULARY REVIEW

SCIENCE PUZZLERS, TWISTERS & TEASERS

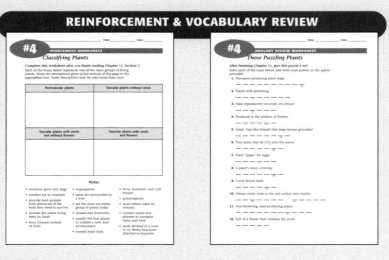

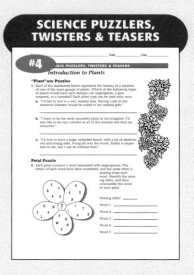

Review & Assessment

STUDY GUIDE

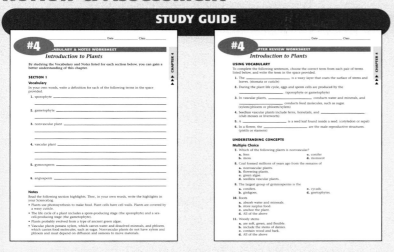

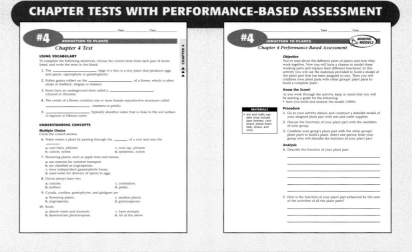

CHAPTER TESTS WITH PERFORMANCE-BASED ASSESSMENT

Lab Worksheets

ECOLABS & FIELD ACTIVITIES

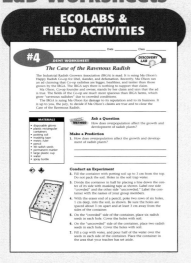

WHIZ-BANG DEMONSTRATIONS

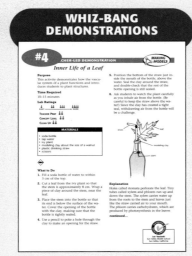

LONG-TERM PROJECTS & RESEARCH IDEAS

DATASHEETS FOR LABBOOK

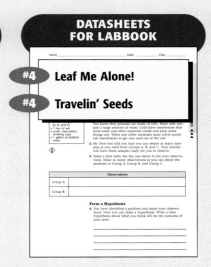

Applications & Extensions

CRITICAL THINKING & PROBLEM SOLVING

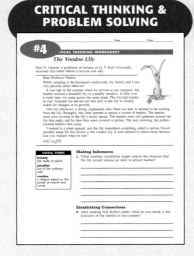

EYE ON THE ENVIRONMENT

INTERACTIVE EXPLORATIONS

SECTION 1

What Makes a Plant a Plant?

▶ Theophrastus

Theophrastus (372–288 B.C.), Aristotle's student, was one of the first botanists. He wrote two books, *History of Plants* and *The Causes of Plants*. He described the morphology, uses, propagation, and pollination of 500 plants and described sexual reproduction in plants. Theophrastus directed the Lyceum (a school and center of learning), in Athens. The Lyceum housed the first botanical garden. Theophrastus's writings were the standard for botanical study until the sixteenth century.

IS THAT A FACT!

- ➥ The leaves of a pitcher plant form tall, narrow cups that hold rainwater. The tip of the plant is colorful and has nectar-secreting glands that attract insects. The insects follow a path of tiny hairs down into the cup, where the walls are smooth. The insects lose their grip and drown.

▶ Carnivorous Plants

Carnivorous plants photosynthesize and are true plants. They tend to grow in bogs and marshes, where the soil is waterlogged. Bacteria and fungi cannot thrive in these soils, so there is little decomposition of organic matter to provide nutrients to plants. The small invertebrates that carnivorous plants catch provide additional nutrients, especially nitrates. The insects are digested by juices secreted by the leaves or by bacteria and fungi living in the plant.

SECTION 2

Seedless Plants

▶ Bryophytes: Good Things Come in Small Packages

Bryophytes, which include mosses, hornworts, and liverworts, make up 15,000 described species worldwide. Various species tend to be restricted to particular environments because of sensitivities to temperature, light exposure, water availability, and chemical composition of the substrate. This makes them good indicator species for applied ecologists and conservation biologists; they can characterize an environment by identifying the bryophytes.

- Bryophytes are also useful for studies of evolution and population genetics. Their structures are complex enough to use as models for land-plant evolution but still simple enough for genetic studies. They are small and are easy to culture in the lab. Field biologists, however, can usually observe them year-round under natural conditions.

▶ Evolution of Ferns

Ferns are an ancient group of plants with fossil records dating to the Devonian period, 408 million years ago. Nearly all of those early fern groups are now extinct. Only one or two living genera can be traced directly to Carboniferous ancestors.

IS THAT A FACT!

- ➥ Before the invention of flashbulbs and strobe lights for indoor and low-light photography, photographers created an explosive flash of light with a powder. They spread the powder on a metal bar attached to a T-shaped device and ignited it. The powder contained millions of spores from club mosses (seedless, vascular plants that are related to ferns and horsetails and are not true mosses).

SECTION 3

Plants with Seeds

▶ The Millennium Seed Bank

The Royal Botanic Gardens, in Kew, England, has launched a project to collect seeds from 25,000 plant species around the world. One-fourth of the world's plants might become extinct in the next 50 years, but the seed bank will ensure the survival of plants vital for stabilizing soil and providing food crops, medicine, and building materials. The collected seeds are dried and stored in subzero temperatures. Scientists believe the seeds will be viable even hundreds of years in the future.

▶ The Economic and Social Importance of Plants

Farming originated thousands of years ago. Three of the earliest cultivated crops—wheat, rice, and corn—today feed more than half of the people in the world.

- Herbs and spices were valued commodities on ancient trade routes. In medieval times, explorers and merchants brought spices to Europe by camel caravan from east Asia.

- Perfume makers use essential oils from a variety of flowers, including rose, orange, lavender, and jasmine.

- The ancient Egyptians made paper from papyrus reeds. Paper can also be made from nettles, bamboo, and other plants. Today, most paper is made from wood pulp.

IS THAT A FACT!

- ◤ Alder fruits rely on water to disperse their seeds. The fruits contain oil droplets to keep the seeds afloat.

SECTION 4

The Structures of Seed Plants

▶ Inflorescence

The cluster of flowers that develops on many plants is called an inflorescence, of which there are two types. In a determinate inflorescence, the peduncle (main axis) terminates in a flower bud that prevents the peduncle from continued growth. In indeterminate inflorescence, the lower buds open first. As the peduncle continues to grow, the youngest flowers are always at the top. There are several forms of indeterminate inflorescence, including the following:

- raceme: each flower of the cluster is on a pedicel (short stem) that extends from the peduncle (snapdragon)

- spike: resembles a raceme but has no pedicels (gladiolus)

- panicle: branched raceme in which each branch has multiple flowers (lilac)

- head: short, dense spike with flowers in a circular mass (dandelion)

- umbel: all the pedicels grow from the same point at the top of the peduncle (onion)

▶ Can't Keep a Good Grass Down

Grass is mowed, eaten, and trampled, and still it grows. Grass buds are not on the ends of the leaf blades but are at ground level. Grazing geese and rabbits, lawn mowers, and people walking do not destroy the buds.

- People eat the seeds of grasses when they consume rice, wheat, oats, barley, rye, corn, millet, and sorghum.

> For background information about teaching strategies and issues, refer to the *Professional Reference for Teachers.*

Introduction to Plants

 Pre-Reading Questions

Students may not know the answers to these questions before reading the chapter, so accept any reasonable response.

Suggested Answers

1. Answers will vary, but students should discuss the function of flowers and fruits in the reproductive cycle of flowering plants.

2. Answers may vary, but students should point out that plants make their own food and that animals have to eat the nutrients they need. Animal cells do not have chlorophyll or cell walls.

Introduction to Plants

Sections

Pre-Reading Questions

1. How do plants use flowers and fruits?
2. How are plants different from animals?

GREEN ALIENS?

In Costa Rica's Monteverde cloud forest, a green pattern begins to unfold. It is hidden from all but the most careful observer. It looks alien, but it is very much of this Earth. It is part of a fern, a plant that grows in moist areas. How do we know this patterned mass is a fern? How do we know a fern is a plant? In this chapter, you will learn what plants are, how they differ from one another, and how they survive and reproduce.

These round clusters, called sori, contain structures that produce spores.

---

internet connect

HRW On-line Resources	**SCiLINKS NSTA**	**Smithsonian Institution**	**CNN fyi.com**
go.hrw.com	**www.scilinks.com**	**www.si.edu/hrw**	**www.cnnfyi.com**
For worksheets and other teaching aids, visit the HRW Web site and type in the keyword: **HSTPL1**	Use the *sci*LINKS numbers at the end of each chapter for additional resources on the **NSTA** Web site.	Visit the Smithsonian Institution Web site for related on-line resources.	Visit the CNN Web site for current events coverage and classroom resources.

OBSERVING PLANT GROWTH

When planting a garden, you bury seeds in the ground, water them, and then wait for tiny sprouts to poke through the soil. What happens to the seeds while they're below the soil? How do seeds grow into plants?

Procedure

1. Fill a **clear 2 L bottle** to within 8 cm of the top with **potting soil.** Your teacher will have already cut off the neck of the bottle.

2. Press **three or four bean seeds** into the soil and against the wall of the bottle. Add an additional 5 cm of potting soil.

3. Cover the sides of the bottle with **aluminum foil** to keep out light. Leave the top uncovered.

4. Water the seeds with about **60 mL of water.** Add more water when the soil dries out.

5. Check on your seeds each day. Record your observations.

Analysis

6. How long did it take for the seeds to germinate?

7. How many seeds grew?

8. Where do the seeds get the energy to start growing?

73

OBSERVING PLANT GROWTH

MATERIALS
FOR EACH GROUP: • 2 L clear-plastic bottle (with top half cut off) • potting soil • bean seeds • aluminum foil • water (60 mL)

Safety Caution

Some students—particularly those who suffer from allergies—may wish to wear protective gloves while handling the soil and seeds. Have students wash their hands when they are finished with the activity.

Teacher's Notes

Cut off the neck of each bottle before distributing it to students.

Make sure students do not cover the seeds with more than 5 cm of potting soil. Doing so will delay the seedling from emerging from the surface.

Be sure students keep seeds evenly moist to ensure germination.

Answers to START-UP Activity

6. Answers will vary. Germination times vary depending on the seeds used. Soaking the seeds in advance will decrease the number of days to germination.

7. Students may report that not all of their seeds germinated.

8. The seed contains stored food molecules that are used for energy.

Focus

What Makes a Plant a Plant?

In this section students will learn the shared characteristics of plants, such as that all plants make their own food, have a cuticle, reproduce with spores and sex cells, and have cells with cell walls. Finally, students will learn that the four main plant groups are classified as either nonvascular or vascular, depending on how materials are transported within the plant.

🔔 Bellringer

Tell students there are four major types of plants. Ask them to try to identify those types and to give at least two examples for each one. Have them review their responses when they have finished reading this section. (Students will likely respond with flowers, trees, weeds, and grasses. They might also list fruits and vegetables. They probably will not classify plants according to the chapter information.)

1) Motivate

DEMONSTRATION

Water Travel in Plants Slice a stalk of celery lengthwise to just below the leaves. Place the two halves in separate beakers, each containing a different color of water. Red and blue food coloring work best. Students should be able to see the veins in the leaves change color after the colored liquids have traveled up the stalk. **Sheltered English**

SECTION 1
READING WARM-UP

Terms to Learn

sporophyte	vascular plant
gametophyte	gymnosperm
nonvascular plant	angiosperm

What You'll Do

- ◆ Identify the characteristics that all plants share.
- ◆ Discuss the origin of plants.
- ◆ Explain how the four main groups of plants differ.

Sugar maple

74

Directed Reading Worksheet Section 1

☑ internet**connect**

SCILINKS
NSTA
TOPIC: Plant Characteristics
GO TO: www.scilinks.org
*sci*LINKS NUMBER: HSTL280

What Makes a Plant a Plant?

Imagine spending a day without anything made from plants. Not only would it be impossible to make chocolate chip cookies, it would be impossible to do many other things, too. You couldn't wear jeans or any clothes made of cotton or linen. You couldn't use any furniture constructed of wood. You couldn't write with wooden pencils or use paper in any form, including money. You couldn't eat anything because almost all food is made from plants or from animals that eat plants. Spending a day without plants would be very hard to do. In fact, life as we know it would be impossible if plants did not exist!

Fern

Plant Characteristics

Plants come in many different shapes and sizes. What do cactuses, water lilies, ferns, and all other plants have in common? Although one plant may seem very different from another, all plants share certain characteristics.

Plants Make Their Own Food One thing that you have probably noticed about plants is that most of them are green. This is because plant cells have chloroplasts. As you learned earlier, chloroplasts are organelles that contain the green pigment *chlorophyll*. Chlorophyll absorbs light energy from the sun. Plants then use this energy to make food molecules, such as glucose. You may recall that this process is called *photosynthesis*.

Plants Have a Cuticle A *cuticle* is a waxy layer that coats the surface of stems, leaves, and other plant parts exposed to air. Most plants live on dry land, and the cuticle is an adaptation that helps keep plants from drying out.

Prickly pear cactus

SCIENCE HUMOR

Q: How many sides are there to a tree?

A: two; inside and outside

Plant Cells Have Cell Walls Plant cells are surrounded by a cell membrane and a rigid cell wall. The cell wall lies outside the cell membrane, as shown in **Figure 1**. The cell wall helps support and protect the plant. Cell walls contain complex carbohydrates and proteins that form a hard material. When the cell reaches its full size, a tough secondary cell wall may develop. Once this wall is formed, a plant cell cannot grow any larger.

Plants Reproduce with Spores and Sex Cells A plant's life cycle can be divided into two parts. Plants spend one part of their lives in the stage that produces spores and the other part in the stage that produces sex cells (egg and sperm cells). The spore-producing stage is called a **sporophyte** (SPOH roh FIET). The stage that produces egg cells and sperm cells is called a **gametophyte** (guh MEET oh FIET). A diagram of the plant life cycle is shown in **Figure 2**.

Spores and sex cells are tiny reproductive cells. Spores that land in a suitable environment, such as damp soil, can grow into new plants. In contrast, sex cells cannot grow directly into new plants. Instead, a male sex cell (sperm cell) must join with a female sex cell (egg cell). The fertilized egg that results may grow into a new plant.

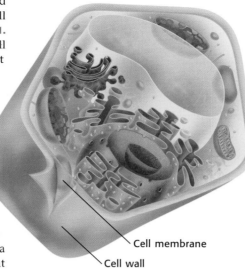

Cell membrane

Cell wall

Figure 1 *In addition to the cell membrane, a cell wall surrounds plant cells.*

Figure 2 Plant Life Cycle

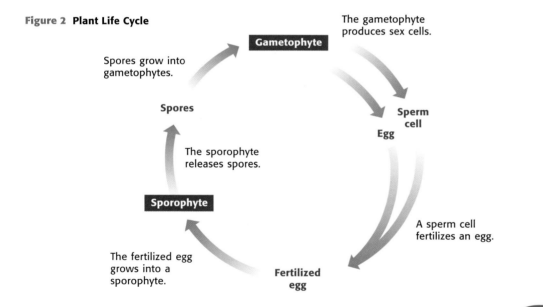

The gametophyte produces sex cells.

Gametophyte

Spores grow into gametophytes.

Spores

Sperm cell

Egg

The sporophyte releases spores.

Sporophyte

A sperm cell fertilizes an egg.

The fertilized egg grows into a sporophyte.

Fertilized egg

75

IS THAT A FACT!

Some flowering plants have a type of surface protector that functions much like sunscreen lotion. Alpine flowers at high elevations produce purple pigments in their leaves to shield them from the damaging effects of ultraviolet light. The pigment allows sunlight in that is needed for photosynthesis but filters out harmful UV rays.

2 Teach

GROUP ACTIVITY

MATERIALS

FOR EACH GROUP:
• baby powder
• water
• eyedroppers or spoons

Organize the class into groups of 3 or 4, and give each group one set of the materials listed above. Tell students that all but one member of the group should coat the palms of their hands with the powder. Instruct the remaining member of the group to release a few drops of water on his or her classmates' hands and to record his or her observations. Explain to students that a plant's cuticle forms a similar barrier to prevent a plant from losing moisture.

Students should wash their hands immediately following this exercise. Sheltered English

MEETING INDIVIDUAL NEEDS

 Advanced Learners
Have students compare and contrast the basic life cycle of a plant with that of bacteria and that of fungi. Tell them to create a series of illustrations that show how all three organism types reproduce and to include captions that explain similarities and differences.

 Teaching Transparency 43 "Plant Life Cycle"

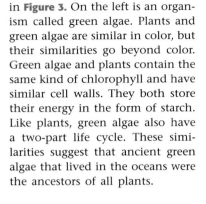

Multicultural CONNECTION

The ancient Maya and Aztec people made extensive use of the breadnut. It was boiled and eaten like potatoes or mashed into a gruel and sweetened. Breadnuts were ground, cooked, and mixed with corn to make tortillas. The diluted sap was fed to babies when their mother's milk was not available. The Maya and Aztecs also fed the leaves to female animals to increase their milk supply.

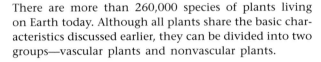

MATH and MORE

Percentages Ask students to suppose the dandelions in their backyard produce 58,791 seeds. The germination rate is 36 percent. How many seeds will germinate?

$(58,791 \times 0.36 = 21,165 \text{ seeds})$

Answer to Self-Check

Plants need a cuticle to keep the leaves from drying out. Algae grow in a wet environment, so they do not need a cuticle.

Teaching Transparency 44 "The Main Groups of Living Plants"

Interactive Explorations CD-ROM "Shut Your Trap!"

internetconnect

SCiLINKS.
NSTA

TOPIC: How Are Plants Classified?
GO TO: www.scilinks.org
sciLINKS NUMBER: HSTL285

The Origin of Plants

If you were to travel back in time 440 million years, Earth would seem like a strange, bare, and unfriendly place. For one thing, no plants lived on land. Where did plants come from?

Green algae

Plant

Take a look at the photographs in **Figure 3.** On the left is an organism called green algae. Plants and green algae are similar in color, but their similarities go beyond color. Green algae and plants contain the same kind of chlorophyll and have similar cell walls. They both store their energy in the form of starch. Like plants, green algae also have a two-part life cycle. These similarities suggest that ancient green algae that lived in the oceans were the ancestors of all plants.

Figure 3 *The similarities between modern green algae and plants suggest that both may have originated from an ancient species of green algae.*

Moss

Liverwort

Figure 4 *Mosses and liverworts are examples of nonvascular plants.*

How Are Plants Classified?

There are more than 260,000 species of plants living on Earth today. Although all plants share the basic characteristics discussed earlier, they can be divided into two groups—vascular plants and nonvascular plants.

Plants Without "Plumbing" The nonvascular plants, mosses and liverworts, are shown in **Figure 4. Nonvascular plants** have no "pipes" to transport water and nutrients. They depend on diffusion and osmosis to move materials from one part of the plant to another. This is possible because nonvascular plants are small. If they were large—the size of trees, for example—there would be no way to deliver the needed materials to all the cells by diffusion and osmosis.

✓ Self-Check

Green algae cells do not have a cuticle surrounding them. Why do plants need a cuticle, while algae do not? *(See page 168 to check your answer.)*

76

IS THAT A FACT!

The coconut palm sends its seed not by air but by sea. Inside the hard-shelled seed, there are supplies for a long voyage, such as plenty of food and water. The shell's exterior has a fibrous coat that helps it float. This self-contained travel package has enabled the coconut to travel miles at sea and to colonize beaches throughout the tropics.

Plants with "Plumbing" Vascular plants do not rely solely on diffusion and osmosis to deliver needed materials to their cells. **Vascular plants** have tissues that deliver needed materials throughout a plant, much as pipes deliver water to faucets in your home. These tissues are called *vascular tissues*. Because vascular tissues can carry needed materials long distances within the plant body, vascular plants can be almost any size.

Vascular plants can be divided into two groups—plants that produce seeds and plants that do not. Plants that do not produce seeds include ferns, horsetails, and club mosses. Plants that produce seeds also fall into two groups—those that produce flowers and those that do not. Nonflowering plants are called **gymnosperms** (JIM noh SPUHRMZ). Flowering plants are called **angiosperms** (AN jee oh SPUHRMZ). The four main groups of living plants are shown in **Figure 5.**

MATH BREAK

Practice with Percents

The following list gives an estimate of the number of species in each plant group:

Mosses
   and liverworts   15,000
Ferns, horsetails,
   and club mosses   12,000
Gymnosperms   760
Angiosperms   235,000

What percentage of plants do not produce seeds?

Figure 5 The Main Groups of Living Plants

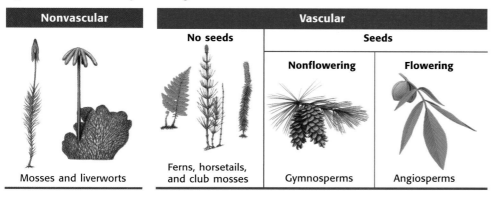

Nonvascular	Vascular		
	No seeds	Seeds	
		Nonflowering	Flowering
Mosses and liverworts	Ferns, horsetails, and club mosses	Gymnosperms	Angiosperms

SECTION REVIEW

1. What are two characteristics that all plants have in common?

2. What type of organism is thought to be the ancestor of all plants? Why?

3. How are ferns, horsetails, and club mosses different from angiosperms?

4. **Applying Concepts** How would you decide whether an unknown organism is a type of green algae or a plant?

internet**connect**

SC*i*LINKS
NSTA

TOPIC: Plant Characteristics, How Are Plants Classified?
GO TO: www.scilinks.org
*sci*LINKS NUMBER: HSTL280, HSTL285

▼ **Answers to Section Review**

1. Plants make their own food, have rigid cell walls, have a cuticle, and reproduce with spores and sex cells.

2. Green algae, because both green algae and plants share many traits including the same kind of chlorophyll.

3. Ferns, horsetails, and club mosses do not produce seeds. Angiosperms do produce seeds.

4. You should determine whether or not the organism has a cuticle. If the organism does, then it is a plant. If the organism does not, then it may be a type of alga.

---

3 Extend

Answer to MATHBREAK

10.3 percent

GROUP ACTIVITY

Writing Arrange a visit to a local plant nursery. Have students record in their ScienceLog at least two examples, with physical descriptions, of each of the three vascular plant groups. Also encourage students to illustrate the plants.

4 Close

Quiz

1. How is a plant's size related to its method of transporting water and nutrients?
(Nonvascular plants rely on osmosis and diffusion, which are efficient only in small plants. Vascular plants have conducting tissues, which enable the plant to be small or very large.)

2. What is required for a spore to grow into a new plant?
(It must land in a suitable environment.)

ALTERNATIVE ASSESSMENT

Have students interview one another about the characteristics common to all plants. For example, have students ask: How do plants make their own food? Why are cell walls necessary? What is the purpose of the cuticle? What are gametophytes and sporophytes, and what roles do they play in the plant's life cycle?

Focus

Seedless Plants

In this section, students will learn that seedless plants include the nonvascular mosses and liverworts and the vascular ferns, horsetails, and club mosses. Students will learn about the features of each group and its life cycles. Finally, students will learn the importance of these plants to the environment and to humans.

 **Bellringer**

Use the board or an overhead projector to display this question:

If plants can make their own food, why do people add fertilizer to the soil? (Fertilizers add nutrients to the soil that plants cannot make for themselves, such as minerals and nitrogen compounds.)

1 Motivate

DISCUSSION

Dandelions Ask students to describe a young dandelion flower and a mature flower that has developed seeds. (The flower's yellow head becomes white and "downy" at summer's end.)

Ask what happens to the "downy" head. (It breaks apart in the wind.)

What is blown away? Explain that although this section is about seedless plants, students can use this visual imagery of the dandelion to understand the different phases of a seedless plant's life cycle.

Terms to Learn

rhizoid
rhizome

What You'll Do

◆ Describe the features of mosses and liverworts.
◆ Describe the features of ferns, horsetails, and club mosses.
◆ Explain how plants without seeds are important to humans and to the environment.

Seedless Plants

Two groups of plants don't make seeds. One group of seedless plants is the nonvascular plants—mosses and liverworts. The other group is made up of several vascular plants—ferns, horsetails, and club mosses.

Mosses and Liverworts

Mosses and liverworts are small. They grow on soil, the bark of trees, and rocks. Because they lack a vascular system, these plants usually live in places that are always wet. Each cell of the plant must absorb water directly from the environment or from a neighboring cell.

Mosses and liverworts don't have true stems, roots, or leaves. They do, however, have structures that carry out the activities of stems, roots, and leaves.

Rock-to-Rock Carpeting Mosses typically live together in large groups, covering soil or rocks with a mat of tiny green plants. Each moss plant has slender, hairlike threads of cells called **rhizoids**. Like roots, rhizoids help hold the plant in place. Each moss plant also has a leafy stalk. The life cycle of the moss alternates between the gametophyte and the sporophyte, as shown in **Figure 6**.

Figure 6 Moss Life Cycle

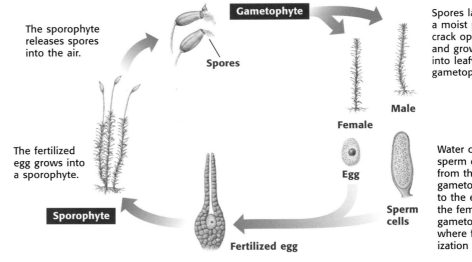

The sporophyte releases spores into the air.

Spores

Gametophyte

Spores land in a moist place, crack open, and grow into leafy gametophytes.

Male

Female

The fertilized egg grows into a sporophyte.

Egg

Sporophyte

Sperm cells

Water carries sperm cells from the male gametophyte to the egg in the female gametophyte, where fertilization occurs.

Fertilized egg

78

CROSS-DISCIPLINARY FOCUS

History Elizabeth Knight Britton was a botanist when female scientists faced many obstacles in their careers. She worked as the unofficial curator of mosses at Columbia University, in New York, and published 346 scientific papers between 1881 and 1930. Britton is credited with being the first person to suggest the establishment of the New York Botanical Garden, located in the Bronx, and she was a founder of the Wild Flower Preservation Society of America. One of her later achievements included helping to enact important conservation laws for the state of New York.

Liverworts Like mosses, liverworts are small, nonvascular plants that usually live in damp or moist places. Liverworts have a life cycle similar to that of mosses. The gametophytes of liverworts can be leafy and mosslike or broad and flattened, like those shown in **Figure 7.** Rhizoids extend out of the lower side of the liverwort body and help anchor the plant.

The Importance of Mosses and Liverworts
Although nonvascular plants are small, they play an important role in the environment. They are usually the first plants to inhabit a new environment, such as newly exposed rock. When the mosses and liverworts die, they form a thin layer of soil in which new plants can grow. New mosses and liverworts cover the soil and help hold it in place. This reduces soil erosion. Mosses also provide nesting materials for birds.

Peat mosses are important to humans. Peat mosses grow in bogs and other wet places. In certain locations, such as Ireland, dead peat mosses have built up thick deposits in bogs. This peat can be taken from the bog, dried, and burned as a fuel.

Ferns, Horsetails, and Club Mosses

Unlike most of their modern descendants, ancient ferns, horsetails, and club mosses grew to be quite tall. The first forests were made up of 40 m high club mosses, 18 m high horsetails, and 8 m high ferns. **Figure 8** shows how these forests may have looked. These plants had vascular systems and could therefore grow taller than nonvascular plants.

Figure 7 *This liverwort has a broad, flattened gametophyte. The sporophyte looks like a tiny palm tree or umbrella.*

Moss Mass

Determine the mass of a small sample of **dry sphagnum moss.** Place this sample in a **large beaker of water** for 10–15 minutes. What do you think the mass will be after soaking in water? Remove the wet moss from the beaker, and determine its mass. How much mass did the moss gain? Compare your findings with your predictions. What could this absorbent plant be used for? Do some research to find out.

Figure 8 *Vascular tissue allowed the ancestors of modern ferns, horsetails, and club mosses to grow tall.*

79

Some of the same factors that encourage the development of peat—high acidity and low oxygen—have preserved the bodies of people who died in bogs. Preserved bodies have been recovered from bogs in Europe and in Florida. Some of the bodies are 2,000 years old.

internetconnect

SC*i*LINKS
NSTA

TOPIC: Seedless Plants
GO TO: www.scilinks.org
*sci*LINKS NUMBER: HSTL290

2 Teach

REAL-WORLD CONNECTION

Sphagnum Moss Sphagnum moss, a primary component of peat bogs, was used during World War I as an absorbent dressing for wounds. Its hollow cells enable it to absorb up to 20 percent of its own weight in water. In earlier times, it was also used for diapers, lamp wicks, and bedding. Today gardeners often used sphagnum moss to protect fragile plants during shipment.

QuickLab

MATERIALS

FOR EACH GROUP:
• dry sphagnum moss
• large beaker of water
• balance or scale
(It may be helpful to use a dry beaker of predetermined mass to hold the wet moss on the scale or balance.)

Answers to QuickLab

Students should subtract the mass of the dry moss from the mass of the wet moss. The gained mass will be the mass of the water that the moss absorbed. See the Real-World Connection above for some uses of sphagnum moss.

MISCONCEPTION ALERT

Folklore says that moss grows on the north side of trees. But it is actually the green alga *Pleurococcus* that thrives on the moist, shaded (usually north) side of trees, stone walls, and fences.

Directed Reading Worksheet Section 2

Writing **Activity** Refer students to the diagram of the fern life cycle. Have them expand the identification of each stage with information they have learned from the text. For example, students could write a physical description of the gametophyte and of the conditions necessary for the sperm cell to fertilize the egg.

ACTIVITY

Identifying Plant Parts Before students read this page, let them view a potted fern. Tell students to draw a picture of the fern and to label as many parts as they can. Ask students if they have drawn and labeled the stem. (Answers will vary. Some students may realize that no stem was visible. Some may label a part of the fronds as a stem.)

Then remove the fern from the soil, point to the rhizome, and ask students if they know what it is. Explain that this structure is an underground stem. Stress to students that the structure and function of a plant part, not its location, is evidence of that part's identity. **Sheltered English**

Homework

Writing **Fern Food** Fiddleheads are a delicacy found in the forests of the northeastern United States during the spring. Edible fiddleheads are actually the tightly coiled, emerging fronds of the fern, *Matteuccia struthiopteris*. Have interested students research recipes for fiddleheads or prepare some dishes for the class.

Ferns Ferns grow in many places, from the cold Arctic to warm, humid tropical forests. Although most ferns are relatively small plants, some tree ferns in the tropics grow as tall as 23 m.

Figure 9 shows a typical fern. Most ferns have an underground stem, called a **rhizome,** that produces leaves called *fronds* and wiry roots. Young fronds are tightly coiled. They are called *fiddleheads* because they look like the end of a violin, or fiddle.

Like the life cycles of all other plants, the life cycle of ferns, shown in **Figure 10,** is divided into two parts. You are probably most familiar with the sporophyte. The fern gametophyte is a tiny plant about the size of half of one of your fingernails. It is green and flat, and it is usually shaped like a tiny heart. The fern gametophyte has male structures that produce sperm cells and female structures that produce eggs. If a thin film of water is on the ground, the sperm cells can swim through it to an egg.

Figure 9 *This fern produces spores in spore cases on the underside of fronds. Fiddleheads grow into new fronds.*

Fiddlehead

Frond

Cluster of spore cases

Figure 10 Fern Life Cycle

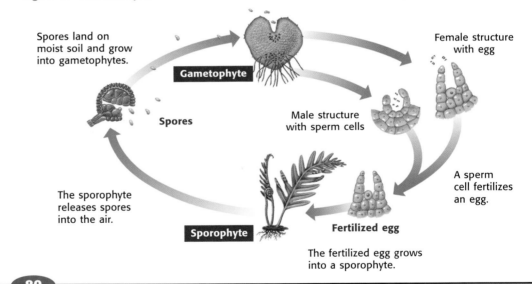

Spores land on moist soil and grow into gametophytes.

Gametophyte

Spores

Female structure with egg

Male structure with sperm cells

A sperm cell fertilizes an egg.

The sporophyte releases spores into the air.

Sporophyte

Fertilized egg

The fertilized egg grows into a sporophyte.

80

IS THAT A FACT!

The largest plant leaves in the world are those of the raffia plant of the Mascarene Islands, in the Indian Ocean, and of the bamboo palm of South America and Africa. These leaves can grow to nearly 20 m in length.

Horsetails Horsetails were common plants millions of years ago, but only about 15 species have survived to the present. Modern horsetails, shown in **Figure 11,** are small vascular plants usually less than 1.3 m tall. They grow in wet, marshy places. Their stems are hollow and contain silica. Because of this, they feel gritty. In fact, pioneers of the early United States called horsetails "scouring rushes" and used them to scrub pots and pans. The life cycle of horsetails is similar to that of ferns.

Club Mosses Club mosses, shown in **Figure 12,** are about 25 cm tall and grow in woodlands. Club mosses are not actually mosses. Unlike true mosses, club mosses have vascular tissue. Like horsetails, club mosses were common plants millions of years ago.

Figure 11 *The conelike tips of horsetails contain spores.*

The Importance of Seedless Vascular Plants Seedless vascular plants play important roles in the environment. Like nonvascular plants, the ferns, horsetails, and club mosses help form soil. They also hold the soil in place, preventing soil erosion.

Ferns are popular as houseplants because of their beautiful leaves. The fiddleheads of some ferns are harvested in early spring, cooked, and eaten.

For humans, some of the most important seedless vascular plants lived and died about 300 million years ago. The remains of these ancient ferns, horsetails, and club mosses formed coal, a fossil fuel that we now extract from the Earth's crust.

Figure 12 *Club mosses release spores from their conelike tips.*

SECTION REVIEW

1. What is the connection between coal and seedless vascular plants?

2. How are horsetails and club mosses similar to ferns?

3. List two ways that seedless vascular plants are important to the environment.

4. **Applying Concepts** Why don't mosses ever grow as large as ferns?

internetconnect

SC*LINKS*
NSTA

TOPIC: Seedless Plants
GO TO: www.scilinks.org
*sci*LINKS NUMBER: HSTL290

81

▼ *Answers to Section Review*

1. Coal is a fossil fuel formed from seedless vascular plants that died about 300 million years ago.

2. Like ferns, horsetails and club mosses are vascular, grow in moist environments, and do not use seeds to reproduce.

3. Seedless vascular plants help to form soil. They also help to prevent erosion.

4. Ferns have vascular tissue that can transport water and nutrients throughout a large plant. Mosses have no vascular tissue, and each cell must get water directly from the environment or from a neighboring cell.

3 Extend

GOING FURTHER

Writing **Botanical Timeline**
Have students construct a timeline that indicates the first appearances of each seedless plant discussed in this section and describes the animal life that existed at the same time. Tell students to include a brief report on which of those animal and plant species exist today.

4 Close

Quiz

1. What's the difference between a rhizoid and a rhizome? (A rhizoid is a threadlike extension of cells that anchors a moss to its substrate. A rhizome is the underground stem of a fern.)

2. Describe the ecological importance of mosses and liverworts. (They can inhabit areas that previously had no plants. When they die, they form a thin soil in which other plants can grow.)

ALTERNATIVE ASSESSMENT

Writing Have students make a chart that organizes information about the differences and similarities of the seedless nonvascular and seedless vascular plants discussed in this section. Have students present their chart to the class.

Focus

Plants with Seeds

Students will learn the characteristics that differentiate plants with seeds from those without seeds and how to compare and contrast seeds and spores. Students will also learn about the physical and reproductive features of gymnosperms and angiosperms as well as the ecological and economic importance of those plants.

🔊 Bellringer

Use the board or an overhead projector to pose this question to students:

If plants cannot move, how do they disperse their seeds?

1 Motivate

ACTIVITY

Seed Types

MATERIALS
FOR EACH CLASS: • grey stripe sunflower seeds (not the small black oil variety)   • pumpkin seeds   • wildflower seed mix (available in nature stores and from catalogs)

Give students all of the seed varieties to examine. Tell students to compare and contrast the seeds in terms of size, shape, color, and texture. Ask students to compare this information with what they know about spores. Then ask students if they think it would be easier to introduce seed plants or seedless plants to a new plot of land. Why? Explain that this section will help them refine their answers. Sheltered English

---

Terms to Learn

pollen
pollination
cotyledon

What You'll Do

- ◆ Compare a seed with a spore.
- ◆ Describe the features of gymnosperms.
- ◆ Describe the features of angiosperms.
- ◆ List the economic and environmental importance of gymnosperms and angiosperms.

Plants with Seeds

Do the plants on this page look familiar to you? They are all seed plants.

As you read earlier, there are two groups of vascular plants that produce seeds—the gymnosperms and the angiosperms. Gymnosperms are trees and shrubs that produce seeds in cones or fleshy structures on stems. Pine, spruce, fir, and ginkgo trees are examples of gymnosperms. Angiosperms, or flowering plants, produce their seeds within a fruit. Peach trees, grasses, oak trees, rose bushes, and buttercups are all examples of angiosperms.

Peaches

Characteristics of Seed Plants

Just like the life cycle of other plants, the life cycle of seed plants alternates between two stages. During part of the cycle, the seed plants are called sporophytes. During another stage, the seed plants are called gametophytes. Gametophytes produce sex cells. But seed plants differ from other plants in the following ways:

- Seed plants produce seeds, structures in which young sporophytes are nourished and protected.

- Unlike the gametophytes of seedless plants, the gametophytes of seed plants do not live independently of the sporophyte. Gametophytes of seed plants are tiny and are always found protected in the reproductive structures of the sporophyte.

- The male gametophytes of seed plants do not need water to travel to the female gametophytes. Male gametophytes develop inside tiny structures that can be transported by the wind or by animals. These dustlike structures are called **pollen.**

These characteristics allow seed plants to live just about anywhere. That is why seed plants are the most common plants on Earth today.

English elm

Desert yucca

82

IS THAT A FACT!

The smallest seeds in the world belong to an epiphytic orchid; 992.25 million of its seeds weigh 1 g.

What's So Great About Seeds?

A seed develops after fertilization takes place. Fertilization is the union of an egg and a sperm cell. A seed is made up of three parts: a young plant (the sporophyte), stored food, and a tough seed coat that surrounds and protects the young plant. These parts are shown in **Figure 13**.

Did you know that seeds like to travel far from home? See how on page 137 of your LabBook.

Figure 13 *A seed contains stored food and a young plant. A seed is surrounded and protected by a seed coat.*

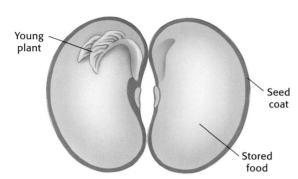

Young plant

Seed coat

Stored food

Plants that reproduce by seeds have several advantages over spore-forming seedless plants. For example, when a seed *germinates,* or begins to grow, the young plant is nourished by the food stored in the seed. By the time the young plant uses up these food reserves, it is able to make all the food it needs by photosynthesis. In contrast, the gametophyte that develops from a spore must be in an environment where it can begin photosynthesis as soon as it begins to grow.

Environment
CONNECTION

Animals need plants to live, but some plants need animals, too. These plants produce seeds with tough seed coats that can't begin to grow into new plants until they have been eaten by an animal. When the seed is exposed to the acids and enzymes of the animal's digestive system, the seed coat wears down. After the seed passes out of the animal's digestive tract, it is able to absorb water, germinate, and grow.

The Accidental Garden

During the summer, Patrick and his sister love to sit out on the porch munching away on juicy watermelon. One year they held a contest to see who could spit the seeds the farthest. The next spring, Patrick noticed some new plants growing in their yard. When he examined them closely, he realized little watermelons were growing on the plants. Patrick and his sister had no idea that they were starting a watermelon garden. Think about the eating habits of animals in the wild. How might they start a garden?

83

2 Teach

PG 137
Travelin' Seeds

GUIDED PRACTICE

Seed Dissection Give each student a lima bean that has been soaked in water to soften its seed coat. Tell students to break apart the bean with their fingers, being careful not to crush or squeeze the bean. Have students compare what they see with the information presented in **Figure 13**. Then tell students to write and illustrate their observations in their ScienceLog.   Sheltered English

Answer to APPLY

Birds, squirrels, and other seed-eating animals scatter some seeds as they eat them or as they eat the fruits that contain them. Sometimes the animals bury seeds to store them, and sometimes the animals excrete the seeds after consuming them in fruit.

Directed Reading Worksheet Section 3

Science Skills Worksheet "Science Writing"

WEIRD SCIENCE

A large number of plants in the heathland of South Africa produce seeds with a very tasty covering called an elaiosome—tasty, that is, to ants, which carry seeds down into their underground colonies. The ants nibble off the outside covering and then leave the seed alone. The ants plant the seed at just the right depth for it to successfully germinate.

Learners Having Difficulty
Many students are already familiar with conifers, such as pine trees, but may suddenly feel confused when the term *gymnosperm* is introduced. Encourage these students to list the characteristics they have observed in pine trees. (stay green all year; thin, needle- or toothpick-like leaves; pine cones)

Stress that while this section provides additional information about gymnosperms, students already know a great deal about them. Sheltered English

CONNECT TO EARTH SCIENCE

Geochronology, the interpretation and dating of the geologic record, includes dendrochronology, which is the study of trees' growth rings. Bristlecone pines have been particularly useful in this endeavor because many of them are very old. Using an increment borer, scientists extract a thin core sample and measure the growth rings from trees of known age. These values are plotted on a graph. Then scientists measure the rings of dead trees of unknown ages. They compare the measurement value of the outer ring of the undated tree with that of a ring in the living tree to find out when the undated tree died. By counting backward from that point, they can determine when the older tree began to grow. Scientists have used this technique to assemble a record that extends back nearly 10,000 years.

Gymnosperms: Seed Plants Without Flowers

The seeds of gymnosperms are not enclosed in a fruit. The word *gymnosperm* is Greek for "naked seed." There are four groups of gymnosperms: conifers, ginkgoes, cycads, and gnetophytes (NEE toh FIETS). Examples are shown in **Figure 14.**

Figure 14 *Gymnosperms do not produce flowers or fruits.*

▲ The **conifers,** with about 550 species, make up the largest group of gymnosperms. Most conifers are evergreen and keep their needle-shaped leaves all year. Conifer seeds develop in cones. Pines, spruces, firs, and hemlocks are examples of conifers.

◄ The **ginkgoes** contain only one living species, the ginkgo tree. Ginkgo seeds are produced in fleshy structures that are attached directly to branches.

▲ The **cycads** were more common millions of years ago. Today there are only about 140 species. These plants grow in the tropics. Like seeds of conifers, seeds of cycads develop in cones.

▲ The **gnetophytes** consist of about 70 species of very unusual plants. This gnetophyte is a shrub that grows in dry areas. Its seeds are formed in cones.

Science Bloopers

When scientists compared radiocarbon dates of bristlecone pines with those obtained from tree ring patterns, they discovered that their calibrations for carbon-14 analysis were incorrect.

The new data indicated that some artifacts found in Europe were 1,000 years older than had been previously thought. The bristlecone pines became known as the "trees that rewrote history."

Gymnosperm Life Cycle The gymnosperms that are most familiar to you are probably the conifers. The name *conifer* comes from Greek and Latin words that mean "carry cones." Conifers have two kinds of cones—male and female. These are shown in **Figure 15.** Male spores are produced in the male cones, and female spores are produced in the female cones. The spores develop into gametophytes. The male gametophytes are pollen, dustlike particles that produce sperm cells. The female gametophyte produces eggs. Wind carries pollen from the male cones to the female cones on the same plant or on different plants. The transfer of pollen is called **pollination.**

After the egg is fertilized, it develops into a seed within the female cone. When the seed is mature, it is released by the cone and falls to the ground. The seed then germinates and grows into a new tree. The life cycle of a pine tree is shown in **Figure 16.**

The Importance of Gymnosperms Conifers are the most economically important group of gymnosperms. People harvest conifers and use the wood for building materials and paper products. Pine trees produce a sticky fluid called resin, which is used to make soap, turpentine, paint, and ink.

Figure 15 *A pine tree has male cones and female cones.*

Figure 16 Pine Life Cycle

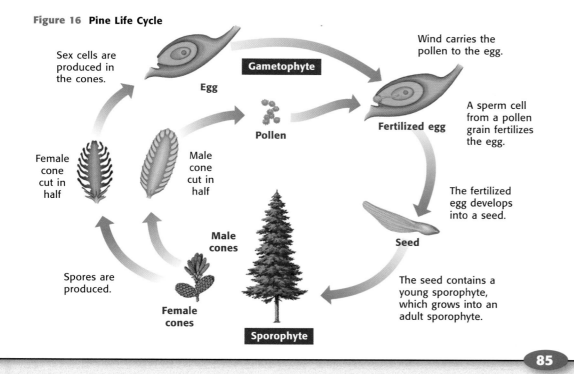

85

People who buy a lot of books on gardening become good weeders.

DISCUSSION

Good Fires Ask students what environmental factors they think are necessary for a gymnosperm to germinate. (sunlight, water, soil)

Tell students that the cones of jack pines must burn to open. Explain that jack pines used to be quite numerous but have been greatly reduced in recent years. Ask students why this has happened. (There is a long-standing national policy of extinguishing forest fires, especially those that may threaten nearby towns and housing developments.)

Tell students that the Kirtland's warbler nests only in young jack pines, and ask what they think has happened to the warbler's population. Explain that conservationists now conduct controlled burns to save both the jack pines and the Kirtland's warblers.

Homework

Writing **Researching Ethnobotany** Have students research the definition of ethnobotany and write a report about ethnobotanists who have studied with indigenous South American groups to learn how they use plants.

*internet***connect**

TOPIC: Plants with Seeds
GO TO: www.scilinks.org
*sci***LINKS NUMBER:** HSTL295

Writing **Art** Before the development of artificial dyes, artists and textile workers used solutions made from berries, roots, bark, leaves, flowers, and seeds of various plants to create colored paints and fabrics. Some of the plants they used included buckthorn, dogwood, fennel, sandalwood, and milkweed. Have students research three natural dyes that are each obtained from a different plant part. Their report should include an illustration of the plant and an explanation of how the dye is processed.

GOING FURTHER

Seed Dispersal

MATERIALS

FOR EACH GROUP:
- cotton balls
- clear or masking tape
- construction paper
- scissors
- fan
- table

Have students shred the cotton balls and place the pieces on a table in front of the fan. Turn on the fan, and have students observe how far the cotton "seeds" travel. Have students wad a strip of tape into a marble-size ball and attach it to the table. How far does the ball move when subjected to the "wind"? Ask students how they think the seeds in this type of fruit might best be transported. (on an animal's fur)

Have students cut a "maple fruit" from construction paper. The rounded tips must be longer on one side than the other. Have students observe how this "fruit" behaves in the wind.

Sheltered English

Angiosperms: Seed Plants with Flowers

Flowering plants, or angiosperms, are the most abundant plants today. Angiosperms can be found in almost every environment on land. There are at least 235,000 species of flowering plants, many more than all other plant species combined. Angiosperms come in a wide variety of sizes and shapes, from dandelions and water lilies to prickly-pear cactuses and oak trees.

All angiosperms are vascular plants that produce flowers and fruits. Tulips and roses are examples of flowering plants with large flowers. Other flowering plants, such as grasses and maple trees, have small flowers. After fertilization, angiosperms produce seeds within fruits. Peaches, lemons, and grapes are fruits, as are tomatoes, cucumbers, and many other foods we think of as vegetables.

What Are Flowers For? Flowers help angiosperms reproduce. Some angiosperms depend on the wind for pollination, but others have flowers that attract animals. As shown in **Figure 17,** when animals visit different flowers, they may carry pollen from flower to flower.

What Are Fruits For? Fruits are also important structures for reproduction in angiosperms. They help to ensure that seeds survive as they are transported to areas where new plants can grow. Fruits surround and protect seeds. Some fruits and seeds, such as those shown in **Figure 18,** have structures that help the wind carry them short or long distances. Other fruits may attract animals that eat the fruits and discard the seeds some distance from the parent plant. Prickly burrs are fruits that are carried from place to place by sticking to the fur of animals or to the clothes and shoes of people.

Figure 17 *This bee is on its way to another squash flower, where it will leave some of the pollen it is carrying.*

Figure 18 *Special structures allow some fruits and seeds to float or drift through the air.*

Dandelion

Maple

Milkweed

IS THAT A FACT!

The giant fan palm, which is native to Seychelles, in Africa, produces a single seed in its fruit that weighs up to 20 kg and takes up to 10 years to develop.

Monocots and Dicots Angiosperms are divided into two classes—monocots and dicots. The two classes differ in the number of cotyledons in their seeds. A **cotyledon** (кант uh LEED uhn) is a seed leaf found inside a seed. Monocot seeds have one cotyledon, and dicot seeds have two cotyledons. Other differences between monocots and dicots are summarized in **Figure 19.** Monocots include grasses, orchids, onions, lilies, and palms. Dicots include roses, cactuses, sunflowers, peanuts, and peas.

The Importance of Angiosperms
Flowering plants provide animals that live on land with the food they need to survive. A deer nibbling on meadow grass is using flowering plants directly as food. An owl that consumes a field mouse is using flowering plants indirectly as food because the field mouse ate seeds and berries.

Humans depend on flowering plants and use them in many ways. All of our major food crops, such as corn, wheat, and rice, are flowering plants. Some flowering plants, such as oak trees, are used to make furniture and toys. Others, such as cotton and flax, supply fibers for clothing and rope. Flowering plants are used to make many medicines as well as cork, rubber, and perfume oils.

Figure 19 Two Classes of Angiosperms

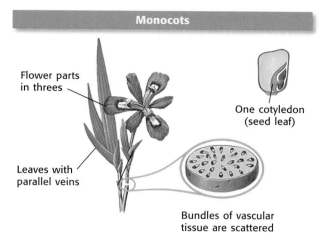

Monocots

Flower parts in threes

One cotyledon (seed leaf)

Leaves with parallel veins

Bundles of vascular tissue are scattered

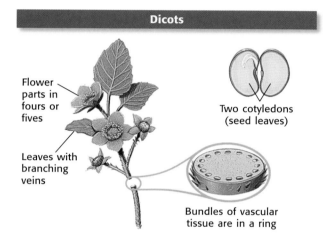

Dicots

Flower parts in fours or fives

Two cotyledons (seed leaves)

Leaves with branching veins

Bundles of vascular tissue are in a ring

SECTION REVIEW

1. What are two differences between a seed and a spore?

2. Briefly describe the four groups of gymnosperms. Which group is the largest and most economically important?

3. How do monocots and dicots differ from each other?

4. **Identifying Relationships** In what ways are flowers and fruits adaptations that help angiosperms reproduce?

internetconnect

SCiLINKS.
NSTA

TOPIC: Plants with Seeds
GO TO: www.scilinks.org
sciLINKS NUMBER: HSTL295

87

▼ *Answers to Section Review*

1. Unlike seeds, spores have no stored food to nourish the new plant. Spores are single-celled. Seeds are made of many cells.

2. Conifers are the most economically important group of gymnosperms. Students should provide descriptions similar to those in **Figure 14.**

3. Students should discuss the differences illustrated in **Figure 19.**

4. The flowers help attract animals that are needed to deliver pollen to other plants. Animals eat the fruit and transport the seeds to areas where new plants can grow.

4) Close

Quiz

1. Give one reason why angiosperms greatly outnumber gymnosperms? (Angiosperms have adapted to almost every environment on land. In some cases, they are helped by animals that carry pollen and disperse seeds.)

2. Which plant group is more important to people, angiosperms or gymnosperms? (Angiosperms; most plant-derived foods for people and animals come from angiosperms.)

ALTERNATIVE ASSESSMENT

Concept Mapping Have students organize the following terms into a concept map:

plants, gymnosperm, angiosperm, flowers, fruits, maple trees, cucumbers, conifers, cycads, monocot, dicot, orchids, lilies, peanuts, saguaro cactus, dandelion

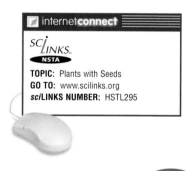 **Teaching Transparency 45** "Two Classes of Angiosperms"

 Reinforcement Worksheet "Classifying Plants"

 Reinforcement Worksheet "Drawing Dicots"

Focus

The Structures of Seed Plants

In this section, students will learn about the physical structures and the functions of a plant's root and shoot systems. Students will also learn how those two factors are related. Finally, students will learn to identify a flower's parts and explain their functions.

 Bellringer

Show students a cactus. Point out the spines, and ask students to identify them and explain their purpose in their ScienceLog. (Spines are modified leaves that help protect the plant from grazing animals.) **Sheltered English**

1) Motivate

DEMONSTRATION

Seed Transport Using a piece of Velcro™, demonstrate how some seeds become dispersed. The burdock plant served as inspiration to the inventor of Velcro. Burdocks are members of the thistle family, and they shed their seeds in burrs. These burrs are transported by animals or in people's clothing to new areas, where they take root. George de Mestral, a Swiss engineer, looked very closely at a burdock burr. He observed that the surface of a burr is covered with tiny hooks. Over the next 8 years, he worked with this design and developed a similar fastener out of nylon. **Sheltered English**

Terms to Learn

xylem	stamen
phloem	pistil
sepal	stigma
petal	ovary

What You'll Do

◆ Describe the functions of roots.
◆ Describe the functions of stems.
◆ Explain how the structure of leaves is related to their function.
◆ Identify the parts of a flower and their functions.

The Structures of Seed Plants

You have different body systems that carry out a variety of functions. For example, your cardiovascular system transports materials throughout your body, and your skeletal system provides support and protection. Similarly, plants have systems too—a root system, a shoot system, and a reproductive system.

Plant Systems

A plant's root system and shoot system supply the plant with needed resources that are found underground and above ground. The root system is made up of roots. The shoot system is made up of stems and leaves.

The root system and the shoot system are dependent on each other. The vascular tissues of the two systems are connected, as shown in **Figure 20**. There are two kinds of vascular tissue—xylem (ZIE luhm) and phloem (FLOH em). **Xylem** transports water and minerals through the plant. **Phloem** transports sugar molecules. Xylem and phloem are found in all parts of vascular plants.

The Root of the Matter

Because most roots are underground, many people do not realize how extensive a plant's root system can be. For example, a 2.5 m tall corn plant can have roots that grow 2.5 m deep and 1.2 m out away from the stem!

Root Functions The main functions of roots are as follows:

- **Roots supply plants with water and dissolved minerals that have been absorbed from the soil.** These materials are transported throughout the plant in the xylem.

- **Roots support and anchor plants.** Roots hold plants securely in the soil.

- **Roots often store surplus food made during photosynthesis.** This food is produced in the leaves and transported as sugar in the phloem to the roots. There the surplus food is usually stored as sugar or starch.

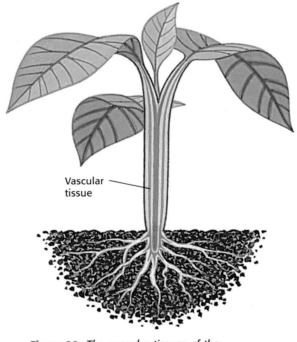

Vascular tissue

Figure 20 *The vascular tissues of the roots and shoots are connected.*

88

Directed Reading Worksheet Section 4

IS THAT A FACT!

The deepest roots ever discovered belonged to a wild fig tree in South Africa. The roots had penetrated to a depth of 122 m.

Root Structure The structures of a root are shown in **Figure 21.** Like the cells in the outermost layer of your skin, the layer of cells that covers the surface of roots is called the *epidermis.* Some cells of the root epidermis extend out from the root. These cells, called *root hairs,* increase the amount of surface area through which roots can absorb water and minerals.

After water and minerals are absorbed by the epidermis, they diffuse into the center of the root, where the vascular tissue is located. Roots grow longer at their tips. A group of cells called the *root cap* protects the tip of a root and produces a slimy substance that makes it easier for the root to grow through soil.

Root Types There are two types of roots—taproots and fibrous roots. Examples of each are shown in **Figure 22.**

A *taproot* consists of one main root that grows downward, with many smaller branch roots coming out of it. Taproots can usually obtain water located deep underground. Dicots and gymnosperms have taproots.

A *fibrous root* has several roots of the same size that spread out from the base of the stem. Fibrous roots typically obtain water that is close to the soil surface. Monocots have fibrous roots.

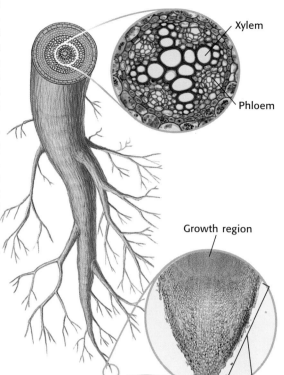

Xylem

Phloem

Growth region

Root cap

Figure 21 *The structures of a root are labeled above.*

Figure 22 *The onion has a fibrous root, and the dandelions and carrots have taproots.*

Fibrous roots

Taproot

Taproots

89

IS THAT A FACT!

Red mangrove trees grow in salt water along tropical coasts. Once the tree is established, additional roots grow down from the branches. After a while, it looks as if the tree's crown is supported by stilts.

2 Teach

DISCUSSION

Attracting Pollinators Discuss in class that flowers serve an important purpose for plants. They are designed to attract pollinators. They also protect pollen and fruit. Ask students to read the section on flowers and to consider how the following factors make flowers successful in the plant world: color, shape, and smell. (The color of a flower may appeal to a certain insect. Bright colors are effective attractors. The shape of a flower may be specifically designed for a specific pollinator. The tubelike flower of a honeysuckle is perfect for the long beak of a hummingbird. The fragrance, or smell, of a flower is designed to attract pollinators; butterflies flock to sweet-smelling flowers, and flies are drawn to rancid-smelling flowers.)

Teaching Transparency 46 "Root Structure"

Science Skills Worksheet "Taking Notes"

ACTIVITY

Have students draw pictures that illustrate the functions of stems listed on this page. Tell them to write captions for each illustration. Sheltered English

CROSS-DISCIPLINARY FOCUS

History The stems of many large aquatic grasses are called *reeds*. After reeds are harvested and dried, they can be used as a resource material to construct many useful products. For thousands of years, arrows, pens, baskets, musical instruments, furniture, and houses have been made out of reeds. Building boats from reeds is an ancient craft and is still practiced in some places where reeds are plentiful. Ancient Egyptian buildings include friezes of ocean-going ships made of reeds.

In the 1960s, a Norwegian explorer named Thor Heyerdahl wondered if a reed ship could have provided people with transportation across the Atlantic Ocean hundreds or even thousands of years ago. To demonstrate that such a journey was possible, he had Bolivian craftsmen build a traditional reed vessel, the *Ra II*, and in 1970, he sailed it across the Atlantic Ocean.

What's the Holdup?

As shown in **Figure 23,** stems vary greatly in shape and size. Stems are usually located above ground, although many plants have underground stems.

Stem Functions A stem connects a plant's roots to its leaves and flowers and performs these main functions:

- **Stems support the plant body.** Leaves are arranged on stems so that each leaf can absorb the sunlight it needs for photosynthesis. Stems hold up flowers and display them to pollinators.

- **Stems transport materials between the root system and the shoot system.** Xylem carries water and dissolved minerals upward from the roots to the leaves and other shoot parts. Phloem carries the glucose produced during photosynthesis to roots and other parts of the plant.

- **Some stems store materials.** For example, the stems of the plants in **Figure 24** are adapted for water storage.

Valley oak

Figure 23 *The stalks of daisies and the trunks of trees are stems.*

Daisy

BRAIN FOOD

Root or shoot? Even though potatoes grow in the ground, they're not roots. The white potato is an underground stem adapted to store starch.

Figure 24 *Baobab trees store large quantities of water and starch in their massive trunks. Cactuses store water in their thick, green stems.*

90

IS THAT A FACT!

Linen is a fabric woven from the long fibers harvested from the stems of flax plants.

WEIRD SCIENCE

Garlic bulbs are fleshy, nutrient-storing stems of garlic plants. Garlic has a pungent flavor and is often used to season foods and as a nutritional supplement. At the annual Garlic Festival, in Gilroy, California, visitors can purchase garlic ice cream and garlic candy.

Stem Structures

Herbaceous Stems

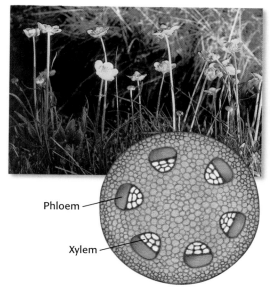

Phloem

Xylem

Woody Stems

Phloem

Xylem

The plants in this group have stems that are soft, thin, and flexible. These stems are called *herbaceous* stems. Examples of plants with herbaceous (her BAY shuhs) stems include wildflowers, such as clovers and poppies, and many vegetable crops, such as beans, tomatoes, and corn. Some plants with herbaceous stems live only 1 or 2 years. A cross section of one kind of herbaceous stem is shown above.

Trees and shrubs have rigid stems made of wood and bark. Their stems are called *woody* stems. If a tree or a shrub lives in an area with cold winters, the plant has a growing period and a dormant period.

At the beginning of each spring growing period, large xylem cells are produced. As fall approaches, the plants produce smaller xylem cells, which appear darker. In the fall and winter, the plants stop producing new cells. The cycle begins again when the spring growing season begins. A ring of dark cells surrounding a ring of light cells make up a growth ring.

SECTION REVIEW

1. What are three functions of roots?

2. What are three functions of stems?

3. **Applying Concepts** Suppose the cross section of a tree reveals 12 light-colored rings and 12 dark-colored rings. How many years of growth are represented?

91

▼ Answers to Section Review

1. Roots supply plants with water and dissolved minerals absorbed from the soil, support and anchor plants, and store surplus food made during photosynthesis.

2. Stems support the plant body, transport materials between the root system and the shoot system, and store materials.

3. 12 years

MEETING INDIVIDUAL NEEDS
Learners Having Difficulty

MATERIALS

FOR EACH CLASS:
- 3 different-colored clays
- butter knife

Have students roll one clay into a thin sheet about 10 × 5 cm. Tell students to roll each of the remaining clays into a thin tube-shaped form about 4 cm long and 0.5 cm in diameter. Have students (1) lay out the clay sheet lengthwise, (2) roll up 3 cm of the sheet, (3) place the first tube on the sheet, (4) roll the sheet another 3 cm, (5) place the second tube on the sheet, and (6) roll the rest of the sheet. Then have students mold the ends of the rolled-up sheet until they form a solid cap at either end. Finally, have an adult cut the clay roll with the butter knife and have the students draw the cross section that is now visible. Sheltered English

ACTIVITY
Analyzing Tree Growth Rings

MATERIALS

FOR EACH GROUP:
- preserved cross sections of trees that clearly show growth rings

Have students identify the xylem and phloem rings. Ask students to measure the width of each ring and to interpret the measurements. Have them take into account the growing conditions in the spring and summer. Sheltered English

Teaching Transparency 47
"Stem Structure"

Leaf Collecting

MATERIALS

FOR EACH GROUP:
- newspapers
- heavy books
- transparent tape
- paper towels
- notebook or 4 × 6 in. index cards.
- leaves

Challenge students to see how many different types of leaves they can collect. They can mount them in a separate notebook or in their ScienceLog. Students should press the leaves a few days after collecting them. They can do this by placing each leaf between two paper towels and stacking heavy books on top. When the leaf is flat and dry, have them tape the leaves to cards or to the pages of a notebook. Have students use reference books to identify the names of the plants. Then have them label each leaf with the common name (such as red oak, honeysuckle, sugar maple), the scientific name, the date, and the location where the leaf was found. Sheltered English

Answer to Self-Check

Stems hold up the leaves so that the leaves can get adequate sunshine for photosynthesis.

Teaching Transparency 48
"Leaf Structure"

A Plant's Food Factories

Leaves vary greatly in shape and size. They may be round, narrow, heart-shaped, or fan-shaped. The raffia palm, shown in **Figure 25,** has leaves that may be six times longer than you are tall. A leaf of the duckweed, a tiny aquatic plant also shown in Figure 25, is so small that several can fit on your fingernail.

Sweet gum

Leaf Function The main function of leaves is to make food for the plant. Leaves capture the energy in sunlight and absorb carbon dioxide from the air. Light energy, carbon dioxide, and water are needed to carry out photosynthesis. During photosynthesis, plants use light energy to make food (sugar) from carbon dioxide and water.

Raffia palm

Mimosa

Figure 25 *Even though the leaves of these plants are very different, they serve the same purpose.*

Duckweed

✓ Self-Check

How is the function of stems related to the function of leaves? *(See page 168 to check your answers.)*

92

IS THAT A FACT!

Duckweed plants, which live on the surfaces of calm ponds, are the smallest flowering plants in the world. They can be less than 1 mm long and weigh about 150 μg, or the equivalent of two grains of table salt. It's a good thing animals like to eat duckweed, because it reproduces exponentially. One species reproduces every 30–36 hours. Left unchecked, one plant could produce 1 nonillion plants (1 followed by 30 zeros) in 4 months!

Leaf Structure The structure of leaves is related to their main function—photosynthesis. **Figure 26** shows a cutaway view of a small block of leaf tissue. The top and bottom surfaces of the leaf are covered with a single layer of cells called the epidermis. Light can easily pass through the thin epidermis to the leaf's interior. Notice the tiny pores in the epidermis. These pores, called *stomata* (singular, *stoma*), allow carbon dioxide to enter the leaf. *Guard cells* open and close the stomata.

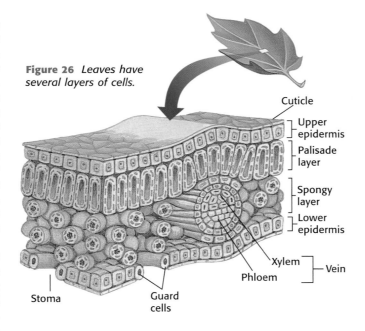

Figure 26 *Leaves have several layers of cells.*

Cuticle
Upper epidermis
Palisade layer
Spongy layer
Lower epidermis
Xylem
Phloem
Vein
Stoma
Guard cells

The middle of a leaf, which is where photosynthesis takes place, has two layers. The upper layer is the *palisade layer,* and the lower layer is the *spongy layer.* Cells in the palisade layer contain many chloroplasts, the green organelles that carry out photosynthesis. Cells in the spongy layer are spread farther apart than cells in the palisade layer. The air spaces between these cells allow carbon dioxide to diffuse more freely throughout the leaf.

The veins of a leaf contain xylem and phloem surrounded by supporting tissue. Xylem transports water and minerals to the leaf. Phloem conducts the sugar made during photosynthesis from the leaf to the rest of the plant.

Leaf Adaptations Some leaves have functions other than photosynthesis. For example, the leaves on a cactus plant are modified as spines. These hard, pointed leaves discourage animals from eating the succulent cactus stem. **Figure 27** shows leaves with a most unusual function. The leaves of a sundew are modified to catch insects. Sundews grow in soil that does not contain enough nitrogen to meet the plants' needs. By catching and digesting insects, a sundew is able to meet its nitrogen requirement.

Figure 27 *This damselfly is trapped in the sticky fluid of a sundew flower.*

RETEACHING

Have students choose a type of plant from the groups presented in this chapter and draw simple illustrations that depict the development of that plant. Tell students to begin with fertilization of the egg by the sperm cell and to continue with germination; the growth of roots, stem, leaves, flowers, and fruits; and seed dispersal; depending on the plant chosen. Sheltered English

INDEPENDENT PRACTICE

Writing A cactus's spines are modified leaves. Other plants have sharp projections to protect them. Roses have thorns, as does the hawthorn tree. Have students research thorns and other "painful" plant parts. Their report should state whether the projection is a leaf modification or another physical structure.

CONNECT TO
PHYSICAL SCIENCE

Photosynthesis is the chemical reaction that provides plants with food from sunlight. Use the following Teaching Transparency to illustrate photosynthesis.

 Teaching Transparency 235
"Photosynthesis"
LINK TO PHYSICAL SCIENCE

IS THAT A FACT!

The more water-repellent a leaf is, the healthier it may be, according to some scientists. Dirt, which contains disease-causing microbes, has a stronger attraction to water droplets than to a leaf's surface. When the water rolls off the leaf, so do the microbes. Water repellence thus helps prevent infection.

RESEARCH

Have students research the flower colors that are most attractive to specific pollinators, such as hummingbirds and bees. (red and yellow, respectively)

Tell students that some pollinators, such as bats, are active at night. Have them research what flower colors best attract nocturnal pollinators. (white)

GOING FURTHER

Life in a Dead Tree
Now that students have learned about live trees, encourage them to explore the life in and around a dead tree, called a snag. Once a snag has fallen down, it is called a log. Have students research the animals that use snags and logs as shelter, including songbirds, hawks, owls, snakes, bats, raccoons, and insects.

Teaching Transparency 48
"Flower Structure"

internet**connect**

SC*i*LINKS.
NSTA

TOPIC: The Structure of Seed Plants
GO TO: www.scilinks.org
*sci*LINKS NUMBER: HSTL300

Flowers

Most people admire the beauty of flowers, such as roses or lilies, without stopping to think about *why* plants have flowers. Flowers are adaptations for sexual reproduction. Flowers come in many different shapes, colors, and fragrances that attract pollinators or catch the wind. Flowers usually contain the following parts: sepals, petals, stamens, and one or more pistils. The flower parts are usually arranged in rings around the central pistil. **Figure 28** shows the parts of a typical flower.

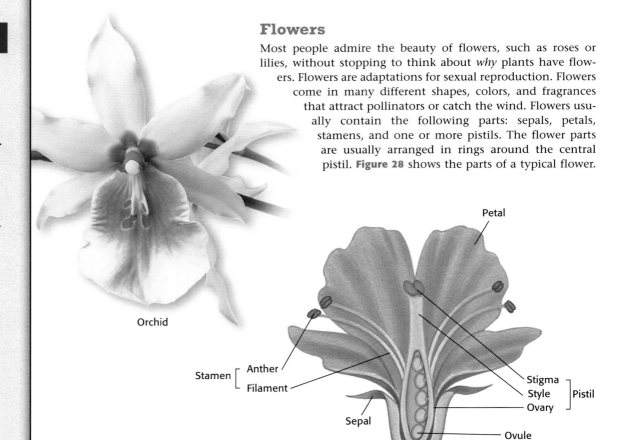

Orchid

Petal

Stamen
Anther
Filament

Stigma
Style
Ovary
Pistil

Sepal

Ovule

Figure 28 *The stamens, which produce pollen, and the pistil, which produces eggs, are surrounded by the petals and the sepals.*

Sepals make up the bottom ring of flower parts. They are often green like leaves. The main function of sepals is to cover and protect the immature flower when it is a bud. As the blossom opens, the sepals fold back so that the petals can enlarge and become visible.

Petals are broad, flat, and thin, like sepals, but they vary in shape and color. Petals may attract insects or other animals to the flower. These animals help plants reproduce by transferring pollen from flower to flower.

IS THAT A FACT!

The heaviest flower in the world is a species of rafflesia from Malaysia, *Rafflesia arnoldii*. The flowers can weigh 11 kg and measure 1 m in diameter.

WEIRD SCIENCE

Carrion-eating flies are probably the pollinators of *Rafflesia arnoldii*. This would explain why the flowers emit a scent reminiscent of rotting meat.

Just above the petals is a circle of **stamens,** which are the male reproductive structures. Each stamen consists of a thin stalk called a *filament,* and each stamen is topped by an *anther.* Anthers are saclike structures that produce pollen grains.

In the center of most flowers is one or more **pistils,** the female reproductive structures. The tip of the pistil is called the **stigma.** Pollen grains collect on stigmas, which are often sticky or feathery. The long, slender part of the pistil is the *style.* The rounded base of the pistil is called the **ovary.** As shown in **Figure 29,** the ovary contains one or more *ovules.* Each ovule contains an egg. If fertilization occurs, the ovule develops into a seed, and the ovary develops into a fruit.

Flowers that have brightly colored petals and aromas usually depend on animals for pollination. Plants without bright colors and aromas, such as the grass flowers shown in **Figure 30,** depend on wind to spread pollen.

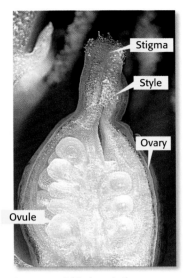

Figure 29 *This hyacinth ovary contains many ovules.*

Figure 30 *The tall stems of these pampas grass flowers allow their pollen to be picked up by the wind.*

SECTION REVIEW

1. Describe the internal structure of a typical leaf. How is a leaf's structure related to its function?

2. Which flower structure produces pollen?

3. **Identifying Relationships** Compare the functions of xylem and phloem in roots, stems, leaves, and flowers.

internet**connect**

SC*i*LINKS.
NSTA

TOPIC: The Structure of Seed Plants
GO TO: www.scilinks.org
*sci*LINKS NUMBER: HSTL300

95

Quiz

1. Why is it that some plants have brightly colored flowers and other plants do not?
(Plants with brightly colored flowers usually need to attract insects, birds, or mammals to ensure pollination. Plants without showy flowers usually rely on the wind to assist with pollination.)

2. Describe the two types of roots, and list at least one plant that has each type.
(A taproot is one main root with smaller branch roots. Dandelions and carrots have a taproot. Fibrous roots are similar in size and grow from the base of the stem. Onions have fibrous roots.)

ALTERNATIVE ASSESSMENT

 *Writing* Have students list ingredients for a salad. (lettuce, tomato, cucumber, carrot, mushroom, red onion, red cabbage, alfalfa sprouts)

Tell them to identify each item as a monocot, dicot, fruit, vegetable, or other. (The mushroom is a fungus.)

Then tell students to list the part of the plant that is eaten. (stem, root, leaves)

Critical Thinking Worksheet
"The Voodoo Lily"

▼ *Answers to Section Review*

1. The epidermis is transparent to allow sunlight in. Stomata allow gases to enter and leave the leaf. The palisade layer in the leaf contains chloroplasts, which carry out photosynthesis. The spongy layer contains cells that are spread out to help the diffusion of carbon dioxide. The veins transport materials to and from the leaf.

2. The anthers produce pollen.

3. Water and nutrients travel through the xylem in the roots, stems, leaves, and flowers. Sugar molecules travel through the phloem from the leaves to other parts of the plant where the sugar is used or stored.

Leaf Me Alone!
Teacher's Notes

Time Required
One 45-minute class period

Lab Ratings

EASY ———————→ HARD

TEACHER PREP   🧪🧪
STUDENT SET-UP   🧪
CONCEPT LEVEL   🧪🧪
CLEAN UP   🧪

MATERIALS

The materials on the student page are enough for a group of 4–5 students. Collect plant specimens ahead of time or have students bring in five specimens that they have collected. Students will need to see the leaves as they appear on the stems, so include as much of the plant as possible.

Safety Caution
Remind students to review all safety cautions and icons before beginning this lab activity.

Lab Notes
This exchange activity can be an effective assessment tool for this lab. Students will enjoy challenging their classmates, and this is a good way for students to learn from each other.

Leaf Me Alone!

Imagine you are a naturalist all alone on an expedition in a rain forest. You have found several plants that you think have never been seen before. You must contact a botanist, a scientist who studies plants, to confirm your suspicion. Because there is no mail service in the rain forest, you must describe these species completely and accurately by radio. The botanist must be able to draw the leaves of the plants from your description.

In this activity, you will carefully describe five plant specimens using the examples and vocabulary lists in this lab.

MATERIALS

- 5 different leaf specimens
- plant guidebook (optional)

Procedure

1. Examine the leaf characteristics illustrated in this lab. These examples can be found on the following page. You will notice that more than one term is needed to completely describe a leaf.

2. In your ScienceLog, draw a diagram of a leaf from each plant specimen.

3. Next to each drawing, carefully describe the leaf. Include general characteristics, such as relative size and color. For each plant, identify the following: leaf shape, stem type, leaf arrangement, leaf edge, vein arrangement, and leaf-base shape. Use the terms and vocabulary lists provided on the next page to describe each leaf as accurately as possible and to label your drawings.

Analysis

4. What is the difference between a simple leaf and a compound leaf?

5. Describe two different vein arrangements in leaves.

6. Based on what you know about adaptation, explain why there are so many different leaf variations.

Going Further
Choose a partner. Using the keys and vocabulary in this lab, describe a leaf, and see if your partner can draw the leaf from your description. Switch roles and see if you can draw a leaf from your partner's description.

96

Science Skills Worksheet
"Using Your Senses"

Jane Lemons
Western Rockingham Middle School
Madison, North Carolina

Leaf Shapes Vocabulary List

cordate—heart shaped

lanceolate—sword shaped

lobate—lobed

oblong—leaves rounded at the tip

orbicular—disk shaped

ovate—oval shaped, widest at base of leaf

peltate—shield shaped

reniform—kidney shaped

sagittate—arrow shaped

Stems Vocabulary List

woody—bark or barklike covering on stem

herbaceous—green, nonwoody stems

Leaf Arrangements Vocabulary List

alternate—alternating leaves or leaflets along stem or petiole

compound—leaf divided into segments or several leaflets on a petiole

opposite—compound leaf with several leaflets arranged oppositely along a petiole

palmate—single leaf with veins arranged around a center point

palmate compound—several leaflets arranged around a center point

petiole—leaf stalk

pinnate—single leaf with veins arranged along a center vein

pinnate compound—several leaflets on either side of a petiole

simple—single leaf attached to stem by a petiole

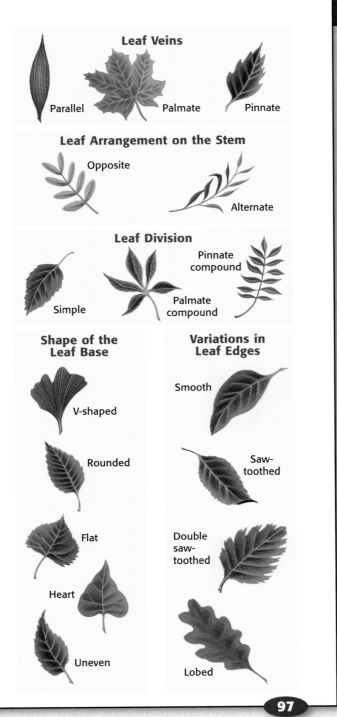

Leaf Veins

Parallel — Palmate — Pinnate

Leaf Arrangement on the Stem

Opposite

Alternate

Leaf Division

Simple — Palmate compound — Pinnate compound

Shape of the Leaf Base

V-shaped

Rounded

Flat

Heart

Uneven

Variations in Leaf Edges

Smooth

Saw-toothed

Double saw-toothed

Lobed

97

Lab Notes

This activity is designed to help students recognize that leaves have many forms and that the variations of leaf shapes are adaptations for a particular function. This exercise also acquaints students with the identification process used in some field or classification guides. Encourage students to recognize the difference between a stem and a petiole and the difference between a leaf and a leaflet.

Answers

4. A simple leaf is a single leaf at the end of a single petiole. A compound leaf has several leaflets on the end of a petiole in various arrangements.

5. Veins can be arranged in three ways: parallel, extending straight up and down the leaf (which is usually elongated); pinnate, which on a single leaf is a network of veins extending from a central vein; and palmate, which is an arrangement of veins originating from a single point on a leaf that has several lobes.

6. Answers will vary but should include adaptations for water conservation, light reception, and insect resistance, among others. Leaves vary because they serve many functions. Explain that leaf shape is only one of the adaptations leaves can have and that the thickness of the leaf and the waxy coating of the leaf are also adaptations. Have students compare an oak leaf with a conifer leaf.

Chapter Highlights

Chapter Highlights

VOCABULARY DEFINITIONS

SECTION 1

sporophyte a stage in a plant's life cycle during which spores are produced

gametophyte a stage in a plant's life cycle during which eggs and sperm are produced

nonvascular plant a plant that depends on the processes of diffusion and osmosis to move materials from one part of the plant to another

vascular plant a plant that has specialized tissues called xylem and phloem, which move materials from one part of the plant to another

gymnosperm a plant that produces seeds but does not produce flowers

angiosperm a plant that produces seeds in flowers

SECTION 2

rhizoids small hairlike threads of cells that help hold nonvascular plants in place

rhizome the underground stem of a fern

SECTION 3

pollen dustlike particles that carry the male gametophytes of seed plants

pollination the transfer of pollen to the female cone in conifers or to the stigma in angiosperms

cotyledon a seed leaf inside a seed

SECTION 1

Vocabulary

sporophyte (p. 75)
gametophyte (p. 75)
nonvascular plant (p. 76)
vascular plant (p. 77)
gymnosperm (p. 77)
angiosperm (p. 77)

Section Notes

- Plants use photosynthesis to make food. Plant cells have cell walls. Plants are covered by a waxy cuticle.

- The life cycle of a plant includes a spore-producing stage (the sporophyte) and a sex-cell-producing stage (the gametophyte).

- Plants probably evolved from a type of ancient green algae.

- Vascular plants have tissues that carry materials throughout a plant. Nonvascular plants do not have vascular tissues and must depend on diffusion and osmosis to move materials.

SECTION 2

Vocabulary

rhizoid (p. 78)
rhizome (p. 80)

Section Notes

- There are two groups of plants that do not make seeds.

- Mosses and liverworts are small, nonvascular plants. They are small because they lack xylem and phloem. Mosses and liverworts need water to transport sperm cells to eggs.

- Ferns, horsetails, and club mosses are vascular plants. They can grow larger than nonvascular plants. Ferns, horsetails, and club mosses need water to transport sperm cells to eggs.

☑ Skills Check

Math Concepts

DO THE PERCENTAGES ADD UP? If 38 percent of the plants in a forest are flowering plants, what percentage of the plants are not flowering plants? The two groups together make up 100 percent. So subtract 38 percent from 100.

100 percent – 38 percent = 62 percent

Look again at the MathBreak on page 77. You can calculate the percentage of plants that do produce seeds by subtracting your MathBreak answer from 100 percent.

Visual Understanding

SEEDS This image shows the two cotyledons of a dicot seed. The seed has been split, and the two cotyledons laid open like two halves of a hamburger bun. You are looking at the inside surfaces of the two cotyledons. Open a peanut and see for yourself. In peanuts, the two cotyledons come apart very easily. You can even see the young delicate plant inside.

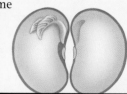

Lab and Activity Highlights

Leaf Me Alone! `PG 96`

Travelin' Seeds `PG 137`

Datasheets for LabBook
(blackline masters for these labs)

SECTION 3

Vocabulary

pollen *(p. 82)*
pollination *(p. 85)*
cotyledon *(p. 87)*

Section Notes

• Seed plants are vascular plants that produce seeds. The sperm cells of seed plants develop inside pollen.

• Gymnosperms are seed plants that produce their seeds in cones or in fleshy structures attached to branches. The four groups of gymnosperms are conifers, ginkgoes, cycads, and gnetophytes.

• Angiosperms are seed plants that produce their seeds in flowers. The two groups of flowering plants are monocots and dicots.

Labs

Travelin' Seeds *(p. 137)*

SECTION 4

Vocabulary

xylem *(p. 88)*
phloem *(p. 88)*
sepal *(p. 94)*
petal *(p. 94)*
stamen *(p. 95)*
pistil *(p. 95)*
stigma *(p. 95)*
ovary *(p. 95)*

Section Notes

• Roots generally grow underground. Roots anchor the plant, absorb water and minerals, and store food.

• Stems connect roots and leaves. Stems support leaves and other structures; transport water, minerals, and food; and store water and food.

• The main function of leaves is photosynthesis. Leaf structure is related to this function.

• Flowers usually have four parts—sepals, petals, stamens, and pistils. Stamens produce sperm cells in pollen. The ovary in the pistil contains ovules. Each ovule contains an egg. Ovules become seeds after fertilization.

SECTION 4

xylem specialized plant tissue that transports water and minerals from one part of the plant to another

phloem specialized plant tissue that transports sugar molecules from one part of the plant to another

sepal a leaflike structure that covers and protects an immature flower

petal the often colorful structure on a flower that is usually involved in attracting pollinators

stamen the male reproductive structure in the flower that consists of a filament topped by a pollen-producing anther

pistil the female reproductive structure in a flower that consists of a stigma, style, and ovary

stigma the flower part that is the tip of the pistil

ovary in flowers, the structure that contains ovules and develops into fruit following fertilization

---

internet connect

GO TO: go.hrw.com

Visit the **HRW** Web site for a variety of learning tools related to this chapter. Just type in the keyword:

KEYWORD: HSTPL1

N S T A

GO TO: www.scilinks.org

Visit the **National Science Teachers Association** on-line Web site for Internet resources related to this chapter. Just type in the *sci*LINKS number for more information about the topic:

TOPIC: Plant Characteristics   *sci*LINKS NUMBER: HSTL280
TOPIC: How Are Plants Classified?   *sci*LINKS NUMBER: HSTL285
TOPIC: Seedless Plants   *sci*LINKS NUMBER: HSTL290
TOPIC: Plants with Seeds   *sci*LINKS NUMBER: HSTL295
TOPIC: The Structure of Seed Plants   *sci*LINKS NUMBER: HSTL300

99

Vocabulary Review Worksheet

Blackline masters of these Chapter Highlights can be found in the **Study Guide.**

Lab and Activity Highlights

LabBank

EcoLabs & Field Activities, The Case of the Ravenous Radish

Whiz-Bang Demonstrations, Inner Life of a Leaf

Long-Term Projects & Research Ideas, Plant Planet

Interactive Explorations CD-ROM

CD 1, Exploration 2, "Shut Your Trap!"

Chapter Review
Answers

USING VOCABULARY

1. cuticle
2. gametophyte
3. xylem/phloem
4. club mosses
5. cotyledon
6. stamens

UNDERSTANDING CONCEPTS

Multiple Choice

7. b
8. d
9. a
10. d
11. c
12. a
13. b
14. c

Short Answer

15. The young plant in a seed is well developed and consists of many cells; a seed can survive in many environments; a seed has stored food for the new plant.
16. Water is necessary for sexual reproduction because sperm cells swim to the egg. Water also opens the spore.

Chapter Review

USING VOCABULARY

To complete the following sentences, choose the correct term from each pair of terms listed below:

1. The __?__ is a waxy layer that coats the surface of stems and leaves. *(stomata* or *cuticle)*

2. During the plant life cycle, eggs and sperm cells are produced by the __?__. *(sporophyte* or *gametophyte)*

3. In vascular plants, __?__ transports water and minerals, and __?__ transports food molecules, such as sugar. *(xylem/phloem* or *phloem/xylem)*

4. Seedless vascular plants include ferns, horsetails, and __?__. *(club mosses* or *liverworts)*

5. A __?__ is a seed leaf found inside a seed. *(cotyledon* or *sepal)*

6. In a flower, the __?__ are the male reproductive structures. *(pistils* or *stamens)*

UNDERSTANDING CONCEPTS

Multiple Choice

7. Which of the following plants is nonvascular?
   a. fern
   b. moss
   c. conifer
   d. monocot

8. Coal formed millions of years ago from the remains of
   a. nonvascular plants.
   b. flowering plants.
   c. green algae.
   d. seedless vascular plants.

9. The largest group of gymnosperms is the
   a. conifers.   c. cycads.
   b. ginkgoes.   d. gnetophytes.

10. Roots
    a. absorb water and minerals.
    b. store surplus food.
    c. anchor the plant.
    d. All of the above

11. Woody stems
    a. are soft, green, and flexible.
    b. include the stems of daisies.
    c. contain wood and bark.
    d. All of the above

12. The veins of a leaf contain
    a. xylem and phloem.
    b. stomata.
    c. epidermis and cuticle.
    d. xylem only.

13. In a flower, petals function to
    a. produce ovules.
    b. attract pollinators.
    c. protect the flower bud.
    d. produce pollen.

14. Monocots have
    a. flower parts in fours or fives.
    b. two cotyledons in the seed.
    c. parallel veins in leaves.
    d. All of the above

Short Answer

15. What advantages does a seed have over a spore?

16. How is water important to the reproduction of mosses and ferns?

Concept Map

17. Use the following terms to create a concept map: nonvascular plants, vascular plants, xylem, phloem, ferns, seeds in cones, plants, gymnosperms, spores, angiosperms, seeds in flowers.

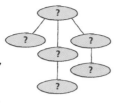

CRITICAL THINKING AND PROBLEM SOLVING

Write one or two sentences to answer the following questions:

18. Plants that are pollinated by wind produce much more pollen than plants that are pollinated by animals. Why do you suppose this is so?

19. If plants did not possess a cuticle, where would they have to live? Why?

20. Grasses do not have strong aromas or bright colors. How might this be related to the way these plants are pollinated?

21. Imagine that a seed and a spore are beginning to grow in a deep, dark crack in a rock. Which reproductive structure—the seed or the spore—is more likely to survive and develop into an adult plant after it begins to grow? Explain your answer.

MATH IN SCIENCE

22. One year a maple tree produced 1,056 seeds. If only 15 percent of those seeds germinated and grew into seedlings, how many seedlings would there be?

INTERPRETING GRAPHICS

23. Examine the cross section of the flower below to answer the following questions:
   a. Which letter corresponds to the structure in which pollen is produced? What is the name of this structure?
   b. Which letter corresponds to the structure that contains ovules? What is the name of this structure?

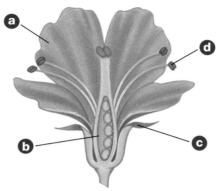

24. In a woody stem, a ring of dark cells and a ring of light cells represent 1 year of growth. Examine the cross section of a tree trunk below, and determine the age of the tree.

Reading Check-up

Take a minute to review your answers to the Pre-Reading Questions found at the bottom of page 72. Have your answers changed? If necessary, revise your answers based on what you have learned since you began this chapter.

101

Concept Map

17. An answer to this exercise can be found at the front of this book.

CRITICAL THINKING AND PROBLEM SOLVING

18. Pollen carried by wind has a smaller probability of reaching another plant of the same species than pollen carried by an animal does. An animal is likely to revisit the same type of plant that it picked up the pollen from.

19. Without a cuticle, plants would lose too much moisture. They might, however, be able to survive in a wet environment.

20. Grasses are not pollinated by insects, so they do not need bright colors or a strong aroma. Their structure is suited to wind pollination.

21. A seed has a better chance of surviving and producing a plant because it has stored nutrients.

MATH IN SCIENCE

22. 158 seedlings

INTERPRETING GRAPHICS

23. a. Pollen is produced in structure *d*, the anther.
   b. Ovules are contained in structure *b*, the ovary.
24. 10 years

Concept Mapping Transparency 12

Blackline masters of this Chapter Review can be found in the **Study Guide.**

Background

The Flavr-Savr™ tomato was approved for consumer use by the Food and Drug Administration in 1994. By inserting an extra gene into the tomato, scientists were able to slow down a series of chemical reactions that cause tomatoes to rot. As a result, this genetically altered tomato can be left to ripen on the vine and can stay on the shelf much longer before becoming soft and spoiling. Many consumers think it also has a better flavor.

Consumer reaction to the tomato was not entirely positive, however. After the FDA's approval was announced, some consumer groups quickly began preparing a boycott of the product as well as planning public "tomato squashes," in which the tomatoes were crushed as a protest against genetically engineered foods.

The tomato marked a new era in commercial agriculture. Genetic engineering may eventually produce crops that can tolerate much harsher growing conditions, including the application of herbicides that are currently too strong to be used.

Science, Technology, and Society

Supersquash or Frankenfruit?

The fruits and vegetables you buy at the supermarket may not be exactly what they seem. Scientists may have genetically altered these foods to make them look and taste better, contain more nutrients, or have a longer shelf life.

From Bullets to Bacteria

Through genetic engineering, scientists are now able to duplicate one organism's DNA and place a certain gene from the DNA into the cells of another species of plant or animal. This technology enables scientists to give plants and animals a new trait. The new trait can then be passed along to the organism's offspring and future generations.

Scientists alter plants by inserting a gene with a certain property into a plant's cells. The DNA is usually inserted by one of two methods. In one method, new DNA is first placed inside a special bacterium, and the bacterium carries the DNA into the plant cell. In another method, microscopic particles of metal coated with the new DNA are fired into the plant cells with a special "gene gun."

High-tech Food

During the past decade, scientists have inserted genes into more than 50 different kinds of plants. In most cases, the new trait from the inserted gene makes the plants more disease resistant or more marketable in some way. For example, scientists have added genes from a caterpillar-attacking bacterium to cotton, tomato, and potato plants. The altered plants produce proteins that kill the crop-eating caterpillars. Scientists are also trying to develop genetically altered peas and red peppers that stay sweeter longer. A genetically altered tomato that lasts longer and tastes better is already in many supermarkets. One day it may even be possible to create a caffeine-free coffee bean.

Are We Ready?

As promising as these genetically engineered foods seem to be, they are not without controversy. Some scientists are afraid that genes introduced to crop plants could be released into the environment or that foods may be changed in ways that endanger human health. For example, could people who are allergic to peanuts become sick from eating a tomato plant that contains certain peanut genes? All of these concerns will have to be addressed before the genetically altered food products are widely accepted.

Find Out for Yourself

▶ Are genetically altered foods controversial in your area? Survey a few people to get their opinions about genetically altered foods. Do they think grocery stores should carry these foods? Why or why not?

◀ A scientist uses a "gene gun" to insert DNA into plant cells.

102

Answer to Find Out for Yourself

Discuss with the class the pros and cons of genetically altered foods. Then ask students to form an opinion about the genetic engineering of animals. Is this any different from that of plants? Why or why not? Accept all reasonable responses. You may wish to divide the class into two teams and organize a debate on the topic of genetically engineered organisms.

CAREERS

ETHNOBOTANIST

Paul Cox is an *ethnobotanist*. He travels to remote places to look for plants that can help cure diseases. He seeks the advice of shamans and other native healers in his search. In 1984, Cox made a trip to Samoa to observe healers. While there he met a 78-year-old Samoan healer named Epenesa. She was able to identify more than 200 medicinal plants, and she astounded Cox with her knowledge. Epenesa had an accurate understanding of human anatomy, and she dispensed medicines with great care and accuracy.

*I*n Samoan culture, the healer is one of the most valued members of the community. After all, the healer has the knowledge to treat diseases. In some cases, Samoan healers know about ancient treatments that Western medicine has yet to discover. Recently, some researchers have turned to Samoan healers to ask them about their medical secrets.

Blending Polynesian and Western Medicine

After Cox spent months observing Epenesa as she treated patients, Epenesa gave him her treatment for yellow fever—a tea made from the wood of a rain-forest tree. Cox brought the yellow-fever remedy to the United States, and in 1986 researchers at the National Cancer Institute (NCI) began studying the plant. They found that the plant contains a virus-fighting chemical called *prostratin.* Further research by NCI indicates that prostratin may also have potential as a treatment for AIDS.

Another compound from Samoan healers treats inflammation. The healers apply the bark of a local tree to the inflamed skin. When a team of scientists evaluated the bark, they found that the healers were absolutely correct. The active compound in the bark, *flavanone,* is now being researched for its medicinal properties. Some day Western doctors may prescribe medicine containing flavanone.

Preserving Their Knowledge

When two of the healers Cox observed in Samoa died in 1993, generations of medical knowledge died with them. The healers' deaths point out the urgency of recording the ancient wisdom before all of the healers are gone. Cox and other ethnobotanists must work hard to gather information from healers as quickly as they can.

The Feel of Natural Healing

▶ The next time you have a mosquito bite or a mild sunburn, consider a treatment that comes from the experience of Native American healers. Aloe vera, another plant product, is found in a variety of lotions and ointments. Find out how well it works for you!

▶ *These plant parts from Samoa may one day be used in medicines to treat a variety of diseases.*

103

CAREERS
Ethnobotanist– Paul Cox

Background

Some biologists estimate that there are 235,000 species of flowering plants in the world. Of these, less than half of 1 percent have been studied for their potential medicinal qualities. Because there are so many species, efficient strategies are necessary to find the plants most likely to have medicinal value.

One strategy used by ethnobotanists is to assume that if native people use a local plant for medicine, then the plant probably has some medicinal value. Many ethnobotanists seek out native healers or shamans. Ethnobotanists hope to acquire the knowledge that has taken the shamans years to accumulate. With these insights, the researchers can then decide which plants they should collect and study.

Some of the most useful drugs developed from plants used by indigenous peoples include aspirin, for reducing pain and inflammation; codeine, for easing pain and suppressing coughs; and quinine, for combating malaria.

Teaching Strategy

Ethnobotanists, like Paul Cox, must be familiar with the names of plants and plant products from many different cultures. Ask students: From what languages have some of our modern English words for plant products been derived?

(There are many examples. *Ginseng* and *tea* came from Chinese, *cinnamon* from Hebrew, *alcohol* and *coffee* from Arabic, *bamboo* from Malay, *pistachio* from Persian, *cashew* from Tupi, and *quinine* from Quechua.)

Chapter Organizer

CHAPTER ORGANIZATION	TIME MINUTES	OBJECTIVES	LABS, INVESTIGATIONS, AND DEMONSTRATIONS	
Chapter Opener pp. 104–105	45	National Standards: UCP 3, SAI 1, SPSP 5, LS 3c, 4c	**Start-Up Activity,** Which End Is Up? p. 105	
Section 1 The Reproduction of Flowering Plants	90	▶ Describe the roles of pollination and fertilization in sexual reproduction. ▶ Describe how fruits are formed from flowers. ▶ Explain the difference between sexual and asexual reproduction in plants. UCP 2–5, SAI 1, 2, SPSP 4, HNS 1, LS 1a, 2a, 2b, 2d, 5b	**Demonstration,** Identify the Parts of a Flower, p. 106 in ATE **QuickLab,** Thirsty Seeds, p. 108 **Labs You Can Eat,** Not Just Another Nut	
Section 2 The Ins and Outs of Making Food	90	▶ Describe the process of photosynthesis. ▶ Discuss the relationship between photosynthesis and cellular respiration. ▶ Explain the importance of stomata in the processes of photosynthesis and transpiration. UCP 1, 5, LS 1a, 1c, 3c, 4c; Labs UCP 1–5, SAI 1, 2, SPSP 3, HNS 2, LS 1a, 4c	**Skill Builder,** Food Factory Waste, p. 118 **Datasheets for LabBook,** Food Factory Waste **Skill Builder,** Weepy Weeds, p. 138 **Datasheets for LabBook,** Weepy Weeds	
Section 3 Plant Responses to the Environment	90	▶ Describe how plants may respond to light and gravity. ▶ Explain how some plants flower in response to night length. ▶ Describe how some plants are adapted to survive cold weather. UCP 1–3, SAI 1, 2, SPSP 2, 3, 5, HNS 1, 2, LS 2b, 2c, 3a, 3c, 3d, 5b	**QuickLab,** Which Way Is Up? p. 114 **Interactive Explorations CD-ROM,** How's It Growing? *A Worksheet is also available in the Interactive Explorations Teacher's Edition.* **EcoLabs & Field Activities,** Recycle! Make Your Own Paper **Long-Term Projects & Research Ideas,** Plant Partners	

TECHNOLOGY RESOURCES

 Guided Reading Audio CD
English or Spanish, Chapter 5

 One-Stop Planner CD-ROM with Test Generator

 Science Discovery Videodiscs
Image and Activity Bank with Lesson Plans: Plant Detectives
Science Sleuths: Green Thumb Plant Rentals #1, Green Thumb Plant Rentals #2

 CNN. Eye on the Environment, Tropical Reforestation, Segment 15

Scientists in Action, Growing Plants in Space, Segment 16

 Interactive Explorations CD-ROM
CD 2, Exploration 8, How's It Growing?

---

*See page **T23** for a complete correlation of this book with the*

NATIONAL SCIENCE EDUCATION STANDARDS.

CLASSROOM WORKSHEETS, TRANSPARENCIES, AND RESOURCES	SCIENCE INTEGRATION AND CONNECTIONS	REVIEW AND ASSESSMENT
Directed Reading Worksheet **Science Puzzlers, Twisters & Teasers** **Science Skills Worksheet,** Using Your Senses		
Transparency 49, Pollination and Fertilization **Transparency 50,** From Flower to Fruit **Directed Reading Worksheet,** Section 1 **Reinforcement Worksheet,** Fertilizing Flowers	**Cross-Disciplinary Focus,** p. 106 in ATE **Cross-Disciplinary Focus,** p. 108 in ATE **Weird Science:** Mutant Mustard, p. 124	**Self-Check,** p. 107 **Section Review,** p. 109 **Quiz,** p. 109 in ATE **Alternative Assessment,** p. 109 in ATE
Transparency 51, Photosynthesis **Directed Reading Worksheet,** Section 2 **Reinforcement Worksheet,** A Leaf's Work Is Never Done **Math Skills for Science Worksheet,** Balancing Chemical Equations **Transparency 256,** Balancing a Chemical Equation **Transparency 52,** Transpiration	**Math and More,** p. 111 in ATE **Connect to Physical Science,** p. 111 in ATE	**Self-Check,** p. 111 **Section Review,** p. 112 **Quiz,** p. 112 in ATE **Alternative Assessment,** p. 112 in ATE
Directed Reading Worksheet, Section 3 **Transparency 53,** Night Length and Blooming Flowers **Transparency 53,** The Change of Seasons and Pigment Color **Reinforcement Worksheet,** How Plants Respond to Change **Critical Thinking Worksheet,** Space Plants	**MathBreak,** Bending by Degrees, p. 113 **Astronomy Connection,** p. 115 **Real-World Connection,** p. 115 in ATE **Apply,** p. 116 **Eye On The Environment:** Rainbow of Cotton, p. 125	**Self-Check,** p. 114 **Homework,** p. 115 in ATE **Section Review,** p. 117 **Quiz,** p. 117 in ATE **Alternative Assessment,** p. 117 in ATE

END-OF-CHAPTER REVIEW AND ASSESSMENT

Chapter Review in Study Guide
Vocabulary and Notes in Study Guide
Chapter Tests with Performance-Based Assessment, Chapter 5 Test
Chapter Tests with Performance-Based Assessment, Performance-Based Assessment 5
Concept Mapping Transparency 13

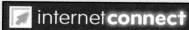

 Holt, Rinehart and Winston On-line Resources
go.hrw.com

For worksheets and other teaching aids related to this chapter, visit the HRW Web site and type in the keyword: **HSTPL2**

 National Science Teachers Association
www.scilinks.org

Encourage students to use the *sci*LINKS numbers listed in the internet connect boxes to access information and resources on the **NSTA** Web site.

Chapter Resources & Worksheets

Visual Resources

TEACHING TRANSPARENCIES

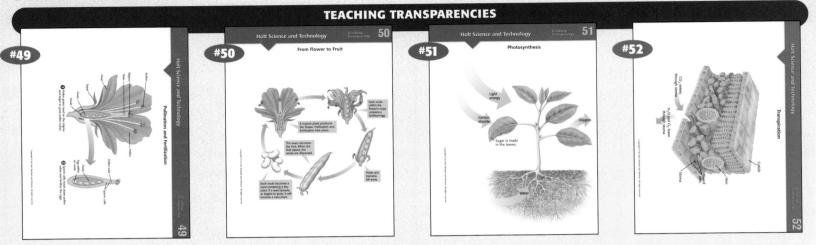

#49 Holt Science and Technology — Pollination and Fertilization — 49

#50 Holt Science and Technology — Teaching Transparency 50 — From Flower to Fruit

#51 Holt Science and Technology — Teaching Transparency 51 — Photosynthesis

#52 Holt Science and Technology — Transpiration — 52

TEACHING TRANSPARENCIES

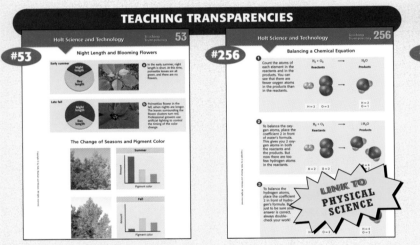

#53 Holt Science and Technology — Teaching Transparency 53 — Night Length and Blooming Flowers / The Change of Seasons and Pigment Color

#256 Holt Science and Technology — Teaching Transparency 256 — Balancing a Chemical Equation

LINK TO PHYSICAL SCIENCE

CONCEPT MAPPING TRANSPARENCY

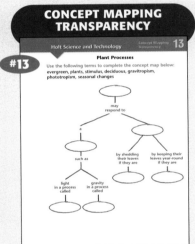

#13 Holt Science and Technology — Concept Mapping Transparency 13 — Plant Processes

Use the following terms to complete the concept map below: evergreen, plants, stimulus, deciduous, gravitropism, phototropism, seasonal changes

Meeting Individual Needs

DIRECTED READING

#5 DIRECTED READING WORKSHEET — Plant Processes

Chapter Introduction
As you begin this chapter, answer the following.
1. Read the title of the chapter. List three things that you already know about this subject.

2. Write two questions about this subject that you would like answered by the time you finish this chapter.

Section 1: The Reproduction of Flowering Plants (p. 106)
3. Which of the following statements is NOT true of flowering plants?
   a. Fertilization takes place within the flower.
   b. They produce seeds in fruit.
   c. They use flowers to reproduce asexually.
   d. They are the largest group of plants in the world.

How Does Fertilization Occur? (p. 106)
4. Flowers can have both male and female reproductive structures.
   True or False? (Circle one.)

REINFORCEMENT & VOCABULARY REVIEW

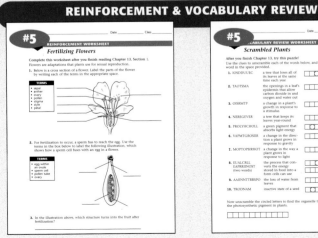

#5 REINFORCEMENT WORKSHEET — Fertilizing Flowers

Complete this worksheet after you finish reading Chapter 13, Section 1.
Flowers are adaptations that plants use for sexual reproduction.
1. Below is a cross section of a flower. Label the parts of the flower by writing each of the terms in the appropriate space.

TERMS
• sepal
• anther
• ovary
• pollen
• stigma
• style
• petal

2. For fertilization to occur, a sperm has to reach the egg. Use the terms in the box below to label the following illustration, which shows how a sperm cell fuses with an egg in a flower.

TERMS
• egg within an ovule
• sperm cell
• pollen tube
• ovary

3. In the illustration above, which structure turns into the fruit after fertilization?

#5 VOCABULARY REVIEW WORKSHEET — Scrambled Plants

After you finish Chapter 13, try this puzzle!
Use the clues to unscramble each of the words below, and write the word in the space provided.
1. IOSDDUUEC — a tree that loses all of its leaves at the same time each year
2. TAOTSMA — the openings in a leaf's epidermis that allow carbon dioxide in and oxygen and water out
3. OISRMTP — a change in a plant's growth in response to a stimulus
4. NEREGEVER — a tree that keeps its leaves year-round
5. PROLYHCHOLL — a green pigment that absorbs light energy
6. VAPMTGROSIIR — a change in the direction a plant grows in response to gravity
7. MOPTOPSIRHOT — a change in the way a plant grows in response to light
8. EUALCRLL EAPRRIENOST (two words) — the process that converts the energy stored in food into a form cells can use
9. AAIINNTTERSPO — the loss of water from leaves
10. TRODNAM — inactive state of a seed

Now unscramble the circled letters to find the organelle that contains the photosynthetic pigment in plants.

SCIENCE PUZZLERS, TWISTERS & TEASERS

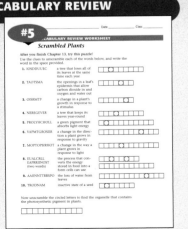
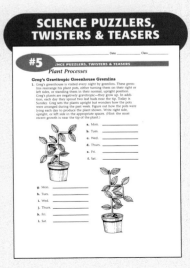

#5 SCIENCE PUZZLERS, TWISTERS & TEASERS — Plant Processes

Greg's Gravitropic Greenhouse Gremlins
1. Greg's greenhouse is visited every night by gremlins. These gremlins rearrange his plant pots, either turning them on their right or left sides, or standing them in their normal, upright position. Greg's plants are negatively gravitropic—they grow up. In addition, each day they sprout two leaf buds near the tip. Today is Sunday. Greg sets the plants upright but wonders how the pots were arranged during the past week. Figure out how the pots were lying each day to produce the plant shown. Write right side, upright, or left side in the appropriate space. (Hint: the most recent growth is near the tip of the plant.)

a. Mon.
b. Tues.
c. Wed.
d. Thurs.
e. Fri.
f. Sat.

g. Mon.
h. Tues.
i. Wed.
j. Thurs.
k. Fri.
l. Sat.

Review & Assessment

STUDY GUIDE

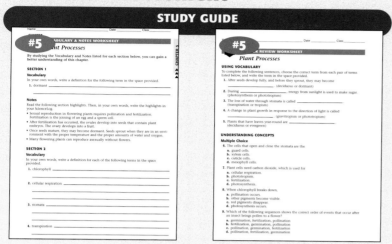

CHAPTER TESTS WITH PERFORMANCE-BASED ASSESSMENT

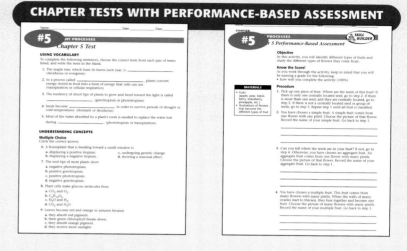

Lab Worksheets

LABS YOU CAN EAT

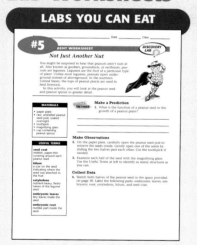

ECOLABS & FIELD ACTIVITIES

LONG-TERM PROJECTS & RESEARCH IDEAS

DATASHEETS FOR LABBOOK

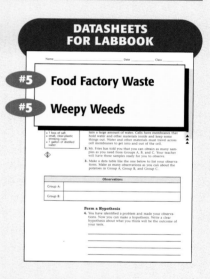

Food Factory Waste

Weepy Weeds

Applications & Extensions

CRITICAL THINKING & PROBLEM SOLVING

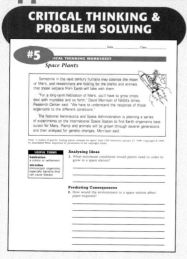

EYE ON THE ENVIRONMENT

SCIENTISTS IN ACTION

INTERACTIVE EXPLORATIONS

The Reproduction of Flowering Plants

▶ Vegetative Reproduction

Vegetative reproduction is another term for *asexual reproduction,* in which a piece of the plant grows into a complete plant. Tulip bulbs produce one or two new bulbs each year, which can be broken off the parent plant and used to produce new plants.

- Succulent plants, such as jade plants, have fleshy leaves full of water. This water sustains the leaves if they fall off the parent plant, often long enough for them to send down roots and develop into new plants.

▶ The Perfect Flower?

Flowers can be either perfect or imperfect. Perfect flowers have both male parts (stamens) and female parts (pistils). Imperfect flowers have one or the other—stamens or pistils—but not both.

IS THAT A FACT!

- ☛ Night-blooming flowers rely on nocturnal animals, such as bats and hawkmoths, to pollinate their flowers. The flowers are usually white for increased visibility and often have a strong fragrance to attract their pollinators.

- ☛ The oldest known fossil seeds are approximately 350 million years old, from the late Devonian period. They belong to plants called seed ferns.

The Ins and Outs of Making Food

▶ Water: A Basic Ingredient

Water conservation is as important for alpine plants as it is for cactuses. At high altitudes there is little rainfall, but there are cold temperatures and high winds. In the Alps, the mountain aven and the mountain kidney vetch have hairlike coverings on their leaves to reduce water loss and provide insulation.

▶ Sunlight

Although sunlight is necessary for photosynthesis, too much sun—specifically, too much ultraviolet radiation—can damage a plant. The sun is more intense on mountains than at lower elevations. Hairlike coverings on leaves can protect some plants from excessive ultraviolet radiation. An example is the silversword, of Hawaii, which grows at altitudes up to 4,000 m.

▶ Clean Air: A Byproduct of Photosynthesis

A study sponsored by NASA has demonstrated that indoor plants in a controlled environment can extract pollutants from the air. The leaves remove low levels of pollutants, and the roots, assisted by activated carbon filters, can remove higher concentrations. The pollutants affected included formaldehyde and benzene gases.

- Scientists continue to study the effectiveness of plants at removing large particle pollutants, such as chemicals from detergents and cleaning fluids; fibers released from clothing, furniture, and insulation; and tobacco smoke.

- Photosynthesis also enables outdoor plants to cleanse the air of carbon emissions from cars and dwellings, nitrogen oxides, airborne ammonia, sulfur dioxide, and ozone.

IS THAT A FACT!

- ☛ One 80 ft beech tree can remove the daily carbon dioxide emissions produced by two single-family homes.

SECTION 3

Plant Responses to the Environment

▶ Plant Pigments

In addition to containing the green pigment chlorophyll, plants contain other pigments that account for the colorful changes in autumn leaves. Xanthophylls are yellow, carotenes are yellowish orange, anthocyanins are red and purple, and tannins are golden yellow.

- In some years, the autumn colors of leaves—especially reds—are bright and colorful; in other years they are dull. Two factors contribute to the difference: warm, sunny days followed by cool nights (temperatures below 45°F) produce bright colors.

IS THAT A FACT!

➡ In 1997, the U.S. Department of Agriculture reported that nearly 60 million poinsettia *(Euphorbia pulcherrima)* plants were purchased in the United States, totaling about $222 million in sales. Poinsettias are the country's most popular potted plants.

▶ Discovery of Auxins

Charles Darwin (1809–1882) is credited with making the first recorded observations that led to the discovery of plant hormones. In 1881, Darwin and his son, Francis, described the occurrence of phototropism, the bending of a plant toward a light source. In an experiment, the Darwins placed lead caps on the growing tips of grass and oat seedlings and noted that the growing tips did not bend. When the lead caps were removed, the tips bent toward the light source. No additional work was done to follow up on the Darwins' experiments until the latter part of the nineteenth century.

▶ Germplasm

A plant's germplasm consists of the plant's genetic material and tissues, organs, and organisms that express the traits contained in the genetic material. The U.S. National Plant Germplasm System (NPGS) conserves and uses germplasm to improve crop plants by transferring disease resistance, improving yields, and adapting plants to grow under new conditions.

IS THAT A FACT!

➡ Plant hormones occur in very small quantities. In a pineapple plant, for example, only 6 μg of auxin are present for 1 kg of plant material. In terms of weight, this is equivalent to a needle in a 20-metric-ton truckload of hay.

For background information about teaching strategies and issues, refer to the *Professional Reference for Teachers.*

Plant Processes

 Pre-Reading Questions

Students may not know the answers to these questions before reading the chapter, so accept any reasonable response.

Suggested Answers

1. Answers will vary, but students may mention wilting, losing leaves, changing colors, and any other changes that show plants respond to the environment.

2. Answers will vary, but students may mention that plants need light in order to stay alive and make food.

Plant Processes

Sections

Pre-Reading Questions

1. How do plants respond to changes in their environment?

2. Why do plants need light?

104

VENUS'S-FLYTRAP

Look at the plant in the photo. Yes, those green spiny pads are its leaves. Why is the Venus's-flytrap such an unusual plant? Unlike most plants, the Venus's-flytrap eats meat. It obtains key nutrients by capturing and digesting insects and other small animals. The two green pads snap shut to trap the prey. In this chapter, you will learn how different types of plants reproduce themselves, respond to their environments, and take in nutrients. You will also learn how plants use sunlight to create food.

---

internet connect

 HRW On-line Resources

 SCiLINKS **NSTA**

 Smithsonian Institution®

 **CNNfyi**.com

go.hrw.com
For worksheets and other teaching aids, visit the HRW Web site and type in the keyword: **HSTPL2**

www.scilinks.com
Use the *sci*LINKS numbers at the end of each chapter for additional resources on the **NSTA** Web site.

www.si.edu/hrw
Visit the Smithsonian Institution Web site for related on-line resources.

www.cnnfyi.com
Visit the CNN Web site for current events coverage and classroom resources.

START-UP Activity

WHICH END IS UP?

If you plant seeds with their "tops" facing in different directions, will their stems all grow upward? Do this activity to find out.

Procedure

1. Pack a **clear medium plastic cup** with slightly **moistened paper towels.**

2. Place **5 or 6 corn seeds,** equally spaced, around the cup between the cup and the paper towels. Point the tip of each seed in a different direction.

3. Use a **marker** to draw arrows on the outside of the cup to indicate the direction that each seed tip points.

4. Place the cup in a well-lit location for one week. Keep the seeds moist by adding **water** to the paper towels as needed.

5. After one week, observe the plant growth. Record the direction in which each plant grew.

Analysis

6. Compare the direction of growth for your seeds. What explanation can you give for the results?

105

START-UP Activity

WHICH END IS UP?

MATERIALS

FOR EACH GROUP:
• clear plastic cup
• paper towels
• water
• corn seeds
• marker

Teacher's Notes

To minimize the effect of light on the growth of the stems, have students rotate the cups every day.

Be sure students keep the paper towels moist throughout the activity.

To further emphasize the effect of gravity on plant growth, have students turn their cups upside down after several days' growth.

Answer to START-UP Activity

6. Students should observe that the stems in all the germinating seeds grow away from the force of gravity.

Focus

The Reproduction of Flowering Plants

This section describes the roles of pollination and fertilization. Students will be able to explain how fruits are formed from flowers and differentiate between sexual and asexual reproduction in flowering plants.

 Bellringer

Ask students to list the names of all the fruits and flowers they can think of in their ScienceLog.

1) Motivate

DEMONSTRATION

Identify the Parts of a Flower
Show students a variety of fresh flowers, and ask them to describe ways the flowers are similar and different. (Student answers should focus on size, structure, fragrance, and color.)

Refer students to **Figure 1.** Point out the anthers, stigmas, petals, and sepals in each flower. Remove the petals, and shake the flower over paper. (Note: Pollen can stain skin and clothing. You may wish to wear protective gloves.)

Ask students to identify the powder on the paper. (pollen)

Explain that pollen contains the flower's male reproductive cells. Sheltered English

 Teaching Transparency 49
"Pollination and Fertilization"

 Teaching Transparency 50
"From Flower to Fruit"

Terms to Learn

dormant

What You'll Do

◆ Describe the roles of pollination and fertilization in sexual reproduction.
◆ Describe how fruits are formed from flowers.
◆ Explain the difference between sexual and asexual reproduction in plants.

The Reproduction of Flowering Plants

If you went outside right now and made a list of all the different kinds of plants you could see, most of the plants on your list would probably be flowering plants. Flowering plants are the largest and the most diverse group of plants in the world. Their success is partly due to their flowers, which are adaptations for sexual reproduction. During sexual reproduction, an egg is fertilized by a sperm cell. In flowering plants, fertilization takes place within the flower and leads to the formation of one or more seeds within a fruit.

How Does Fertilization Occur?

In order for fertilization to occur, sperm cells must be able to reach eggs. The sperm cells of a flowering plant are contained in pollen grains. Pollination occurs when pollen grains are transported from anthers to stigmas. This is the beginning of fertilization, as shown in **Figure 1.** After the pollen lands on the stigma, a tube grows from the pollen grain through the style to the ovary. Inside the ovary are ovules. Each ovule contains an egg.

Sperm cells within the pollen grain move down the pollen tube and into an ovule. Fertilization occurs as one of the sperm cells fuses with the egg inside the ovule.

Figure 1 *Fertilization occurs after pollination.*

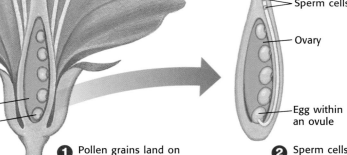

Anther
Pollen
Stigma
Style
Petal
Sepal
Ovary
Ovule

Pollen tube
Sperm cells
Ovary
Egg within an ovule

❶ Pollen grains land on the stigma and begin to grow pollen tubes.

❷ Sperm cells travel down pollen tubes and fertilize the eggs.

106

CROSS-DISCIPLINARY FOCUS

History In 1519, the explorer Hernando Cortez brought cacao beans and a recipe from Montezuma's court back to Spain. The recipe was for a new drink called chocolate made from the beans of the cacao plant. Served with honey and sugar, it greatly impressed the members of the Spanish court. In fact, they were so impressed that they kept this wonderful bean a secret for a hundred years. For the next century, Spanish monks were the only people in Europe who knew how to prepare chocolate.

From Flower to Fruit

After fertilization takes place, the ovule develops into a seed that contains a tiny, undeveloped plant. The ovary surrounding the ovule develops into a fruit. **Figure 2** shows how the ovary and ovules of a flower develop into a fruit and seeds.

Figure 2 *Fertilization leads to the development of fruit and seeds.*

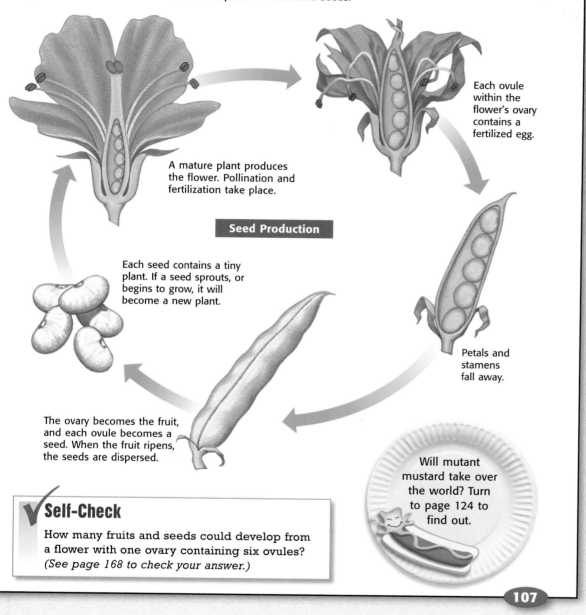

A mature plant produces the flower. Pollination and fertilization take place.

Each ovule within the flower's ovary contains a fertilized egg.

Seed Production

Each seed contains a tiny plant. If a seed sprouts, or begins to grow, it will become a new plant.

Petals and stamens fall away.

The ovary becomes the fruit, and each ovule becomes a seed. When the fruit ripens, the seeds are dispersed.

Will mutant mustard take over the world? Turn to page 124 to find out.

✓ Self-Check

How many fruits and seeds could develop from a flower with one ovary containing six ovules? *(See page 168 to check your answer.)*

107

Directed Reading Worksheet Section 1

internetconnect

SCiLINKS
NSTA

TOPIC: Reproduction of Plants
GO TO: www.scilinks.org
*sci*LINKS NUMBER: HSTL305

2 Teach

ACTIVITY

Germination

MATERIALS
FOR EACH GROUP:
• packet of bean seeds
• two small, plastic containers with snap-on caps
• water

Show students that plant germination can push caps off bottles. This demonstration will take a few days. Fill a container with beans. Fill another container with beans and water. Snap the caps onto the bottles. (Don't use child-proof bottles that lock.) Place them where they can be observed. In a few days, the germinating beans will knock the caps off or even split the bottle apart. Ask students to note which beans are more powerful, the beans with water or the beans without water. (The beans with water are producing a gas, and they are also expanding as they germinate.) **Sheltered English**

GROUP ACTIVITY

Concept Mapping Have students work together in groups of three or four to create a concept map that details the process of sexual reproduction in plants from the time the pollen grains reach the stigma until a seed develops inside a fruit. Encourage students to illustrate their work.

Answer to Self-Check

Fruit develops from the ovary, so the flower can have only one fruit. Seeds develop from the ovules, so there should be six seeds.

CROSS-DISCIPLINARY FOCUS

Art Have students use books from the library or from their home to find examples of how artists (painters, sculptors, photographers, and so on) throughout history have represented flowers and fruits in their work. You might want to show students reproductions of works by Claude Monet or Georgia O'Keeffe to encourage their interest in this activity. Interested students could create a drawing of a flower or a still life of fruit. Sheltered English

MATERIALS

FOR EACH GROUP:
• 12 bean seeds
• 2 Petri dishes
• marking pen or wax pencil
• water

Students should observe that the bean seeds that soaked in water increased in size. Students might also observe that the soaked seeds have begun to crack open.

Answer to QuickLab

5. The seeds swell because they are absorbing water. The stored water can then be used by the young plant.

Reinforcement Worksheet
"Fertilizing Flowers"

Familiar Fruits

While the ovules are developing into seeds, the ovary is developing into the fruit. As the fruit swells and ripens, it holds and protects the developing seeds. Look below to see which parts of the fruits developed from a flower's ovary and ovules.

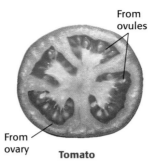

From ovules

From ovary **Tomato**

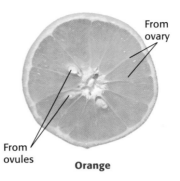

From ovary

From ovules **Orange**

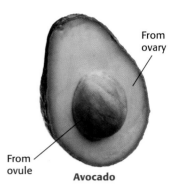

From ovary

From ovule **Avocado**

Seeds Become New Plants

Once a seed is fully developed, the young plant inside the seed stops growing. The seed may become dormant. When seeds are **dormant,** they are inactive. Dormant seeds can often survive long periods of drought or freezing temperatures. Some seeds need extreme conditions such as cold winters or even forest fires to break their dormancy.

When a seed is dropped or planted in an environment that has water, oxygen, and a suitable temperature, the seed sprouts. Each plant species has an ideal temperature at which most of its seeds will begin to grow. For most plants, the ideal temperature for growth is about 27°C (80.6°F). **Figure 3** shows the *germination,* or sprouting, of a bean seed and the early stages of growth in a young bean plant.

QuickLab

Thirsty Seeds

1. Obtain **12 dry bean seeds, 2 Petri dishes,** a **wax pencil,** and **water** from your teacher.

2. Fill one Petri dish two-thirds full of water and add six seeds. Label the dish "Water."

3. Add the remaining seeds to the dry Petri dish. Label this dish "Control."

4. The next day, compare the size of the two sets of seeds. Write your observations in your ScienceLog.

5. What caused the size of the seeds to change? Why might this be important to the seed's survival?

Figure 3 *Sexual reproduction produces seeds that grow into new plants.*

108

WEIRD SCIENCE

In 1967, the seeds of the arctic tundra lupine (*Lupinus arcticus*) were discovered by a mining engineer in a frozen animal burrow in the Yukon. Carbon dating showed the animal remains in the burrow to be at least 10,000 years old. Frozen seeds of the tundra lupine were found among the animal remains. When tested, these seeds germinated within 48 hours!

Other Methods of Reproduction

Many flowering plants can also reproduce asexually. Asexual reproduction in plants does not involve the formation of flowers, seeds, and fruits. In asexual reproduction, a part of a plant, such as a stem or root, produces a new plant. Several examples of asexual reproduction are shown below in **Figure 4**.

Figure 4 *Asexual reproduction can occur in several different ways. Some examples are shown here.*

▼ The strawberry plant produces runners, stems that run horizontally along the ground. Buds along runners grow into new plants.

▲ The "eyes" of potatoes are buds that can grow asexually into new plants.

Kalanchoe plants produce plantlets, ▶ tiny plants along the margins of their leaves. Plantlets eventually fall off and root in the soil as separate plants.

SECTION REVIEW

1. How does pollination differ from fertilization?

2. Which part of a flower develops into a fruit?

3. **Relating Concepts** What do flowers and runners have in common? How are they different?

4. **Identifying Relationships** When might asexual reproduction be important for the survival of some flowering plants?

---

internetconnect

*sci*LINKS
NSTA

TOPIC: Reproduction of Plants
GO TO: www.scilinks.org
*sci*LINKS NUMBER: HSTL305

109

▼ **Answers to Section Review**

1. Pollination occurs when the pollen lands on the stigma of the female flower. Sperm cells from the pollen grain move down into the ovary. Fertilization occurs when one sperm cell fuses with the egg inside the ovule.

2. The ovary develops into the fruit.

3. They both lead to the formation of new plants, but the runner is a form of asexual reproduction while the flower is part of sexual reproduction.

4. Asexual reproduction becomes important when conditions are unfavorable for sexual reproduction.

RESEARCH

Writing Have students research the following different methods by which plants reproduce asexually: corm (gladiolus and crocus), tuber, (gloxinia and potato), tuberous root (dahlia), and rhizome or rootstock (iris and canna). Have students prepare an illustrated report to share with the class.

PORTFOLIO

4 **Close**

Quiz

1. What are the parts of a pistil? (the stigma, style, and ovary) What are the "male" parts of a flower? (the anther and pollen)

2. What is the difference between sexual and asexual reproduction in flowering plants? (Sexual reproduction involves the joining of sperm cell and egg—fertilization—to form a seed that can grow and develop into a new plant. Asexual reproduction does not involve the formation of flowers, seeds, and fruits; instead, a new plant grows from a part—such as a stem or root—of an existing plant.)

ALTERNATIVE ASSESSMENT

Have students make drawings of two common flowers in their ScienceLog and label the petals, sepals, anthers, and stigmas.
Sheltered English

Focus

The Ins and Outs of Making Food

This section describes photosynthesis. Students will learn about the relationship between photosynthesis and cellular respiration. In addition, they will learn about the importance of stomata in the processes of photosynthesis and transpiration.

Bellringer

Write the following question on the board or overhead projector:

> Where do you get the energy you need to stay alive?

Have students write and explain their answer in their ScienceLog. (Students will probably answer that they get their energy from the foods they eat.)

Sheltered English

1 Motivate

DISCUSSION

Food and Energy Ask students the following questions before they begin reading this section.

- What kinds of foods do you eat? (Answers will vary.)
- Where do the animals you eat for food get their energy to survive? (Animals eat plants or other animals that eat plants.)
- Where do plants get their energy to survive? (Plants get their energy from sunlight.)

Explain to students that plants use energy from the sun to make their own food in a process called photosynthesis.

SECTION 2
READING WARM-UP

Terms to Learn

chlorophyll
cellular respiration
stomata
transpiration

What You'll Do

- ◆ Describe the process of photosynthesis.
- ◆ Discuss the relationship between photosynthesis and cellular respiration.
- ◆ Explain the importance of stomata in the processes of photosynthesis and transpiration.

The Ins and Outs of Making Food

Plants do not have lungs, but they need air just like you. Air is a mixture of oxygen, carbon dioxide, and other gases. Plants must have carbon dioxide to carry out photosynthesis, which is how they make their own food.

What Happens During Photosynthesis?

Plants need sunlight to produce food. During photosynthesis, the energy in sunlight is used to make food in the form of the sugar glucose ($C_6H_{12}O_6$) from carbon dioxide (CO_2) and water (H_2O). How does this happen?

Capturing Light Energy Plant cells have organelles called chloroplasts. Chloroplasts contain **chlorophyll,** a green pigment that absorbs light energy. You may not know it, but sunlight is actually a mixture of all the colors of the rainbow. **Figure 5** shows how light from the sun can be separated into different colors when passed through a triangular piece of glass called a prism. Chlorophyll absorbs all of the colors in light except green. Plants look green because chlorophyll reflects green light.

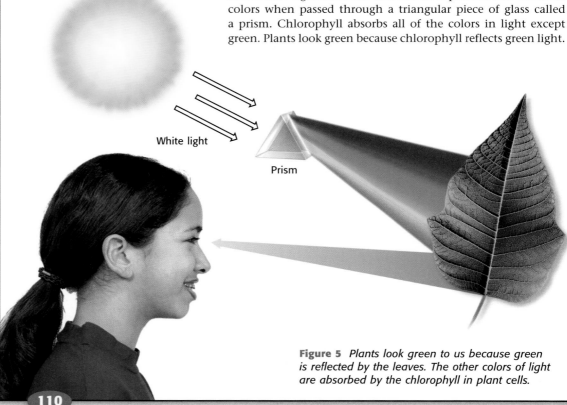

White light

Prism

Figure 5 *Plants look green to us because green is reflected by the leaves. The other colors of light are absorbed by the chlorophyll in plant cells.*

 Teaching Transparency 51 "Photosynthesis"

 Directed Reading Worksheet Section 2

 Reinforcement Worksheet "A Leaf's Work Is Never Done"

MISCONCEPTION ALERT

Students may think that photosynthesis occurs only in the leaves of plants. In some plants, such as aspen trees and cactuses, photosynthesis occurs in the plants' trunks or stems.

Making Sugar The light energy absorbed by chlorophyll is used to break water (H_2O) down into hydrogen (H) and oxygen (O). The hydrogen is then combined with carbon dioxide (CO_2) from the air to make a sugar called glucose ($C_6H_{12}O_6$). Oxygen is given off as a byproduct. The process of photosynthesis is summarized in the following chemical equation:

$$6CO_2 + 6H_2O \xrightarrow{\text{light energy}} C_6H_{12}O_6 + 6O_2$$

The equation shows that it takes six molecules of carbon dioxide and six molecules of water to produce one molecule of glucose and six molecules of oxygen. The process is illustrated in **Figure 6.**

The energy stored in food molecules is used by plant cells to carry out their life processes. Within each living cell, glucose and other food molecules are broken down in a process called cellular respiration. **Cellular respiration** converts the energy stored in food into a form of energy that cells can use. During this process, the plant uses oxygen and releases carbon dioxide and water.

> ✓ **Self-Check**
>
> What is the original source of the energy stored in the sugar produced by plant cells? *(See page 168 to check your answer.)*

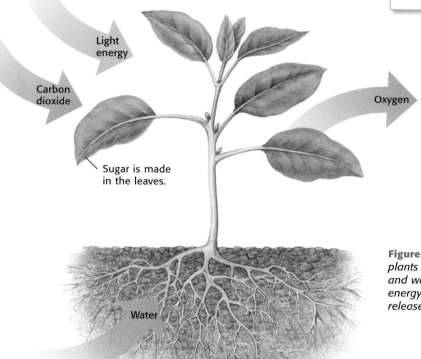

 Light energy

Carbon dioxide

Oxygen

Sugar is made in the leaves.

Water

Figure 6 *During photosynthesis, plants take in carbon dioxide and water and absorb light energy. They make sugar and release oxygen.*

111

Answer to Self-Check

The sun is the source of the energy in sugar.

📄 **internet connect**

SCiLINKS
NSTA

TOPIC: Photosynthesis
GO TO: www.scilinks.org
sciLINKS NUMBER: HSTL310

MEETING INDIVIDUAL NEEDS

Learners Having Difficulty
Have students with limited English proficiency read the definitions of *photosynthesis, cellular respiration,* and *transpiration* and then write a definition of each term in their own words.
Sheltered English

MATH and MORE

Write the following equation— with missing numbers—for photosynthesis on the board:

$6CO_2 + \_\_H_2O + \text{light energy} \rightarrow C\_\_H_{12}O_6 + \_\_O_2$

Explain to students that the same number of atoms of each element must appear on each side of the equation. Have students provide the missing numbers. (6, 6, 6)

 Math Skills Worksheet
"Balancing Chemical Equations"

CONNECT TO PHYSICAL SCIENCE

Use the following Teaching Transparency to help students understand the process of balancing chemical equations.

 Teaching Transparency 256
"Balancing a Chemical Equation"
LINK TO PHYSICAL SCIENCE

LabBook PG 138
Weepy Weeds

GROUP ACTIVITY

Divide students into groups of four. Provide each group with 6 black marbles representing carbon atoms, 18 white marbles representing oxygen, and 12 blue marbles representing hydrogen. Have students arrange the marbles to demonstrate that six molecules of carbon dioxide and six molecules of water are needed to produce one molecule of sugar and six molecules of oxygen gas. **Sheltered English**

4 Close

Quiz

1. What molecules do plants use to make sugar? (carbon dioxide and water)

2. What substances enter and exit a leaf through the stomata? (enter: carbon dioxide; exit: oxygen and water)

ALTERNATIVE ASSESSMENT

Writing Have students write two paragraphs about stomata in their ScienceLog. The first paragraph should describe the appearance of stomata and the passage of materials through the stomata when light is available. The second paragraph should describe the same two events when it is dark.

Teaching Transparency 52
"Transpiration"

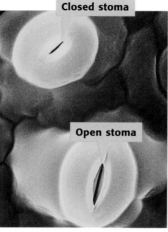

Closed stoma

Open stoma

Figure 7 *When light is available for photosynthesis, the stomata are usually open. When it's dark, the stomata close to conserve water.*

Gas Exchange

All above ground plant surfaces are covered by a waxy cuticle. How does a plant obtain carbon dioxide through this barrier? Carbon dioxide enters the plant's leaves through the **stomata** (singular, *stoma*). A stoma is an opening in the leaf's epidermis and cuticle. Each stoma is surrounded by two *guard cells,* which act like double doors, opening and closing the gap. You can see open and closed stomata in **Figure 7.** The function of stomata is shown in **Figure 8.**

When the stomata are open, carbon dioxide diffuses into the leaf. The oxygen produced during photosynthesis diffuses out of the leaf cells and exits the leaf through the stomata.

When the stomata are open, water vapor also exits the leaf. The loss of water from leaves is called **transpiration.** Most of the water absorbed by a plant's roots is needed to replace water lost during transpiration. When a plant wilts, it is usually because more water is being lost through its leaves than is being absorbed by its roots.

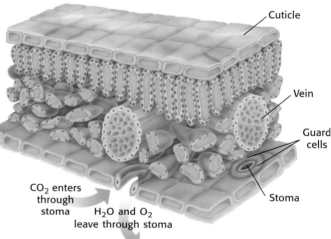

Cuticle

Vein

Guard cells

Stoma

CO$_2$ enters through stoma

H$_2$O and O$_2$ leave through stoma

Figure 8 *Plants take in carbon dioxide and release oxygen and water through the stomata in their leaves.*

internetconnect

SCI**LINKS**
NSTA

TOPIC: Photosynthesis
GO TO: www.scilinks.org
*sci***LINKS NUMBER:** HSTL310

SECTION REVIEW

1. What three things do plants need to carry out photosynthesis?

2. Why must plant cells carry out cellular respiration?

3. **Identifying Relationships** How are the opening and closing of stomata related to transpiration? When does transpiration occur?

▼ *Answers to Section Review*

1. Plants need light, water, and carbon dioxide for photosynthesis.

2. because the energy stored in food must be converted into a form of energy that cells can use

3. Transpiration occurs when water vapor exits the leaf through the stomata. It usually occurs when there is light available for photosynthesis and gas exchange is occurring.

Terms to Learn

tropism   evergreen
phototropism   deciduous
gravitropism

What You'll Do

- Describe how plants may respond to light and gravity.
- Explain how some plants flower in response to night length.
- Describe how some plants are adapted to survive cold weather.

$$\div \; 5 \; \div \; \overset{\Omega}{_{+}} \; \leq \; \infty \; +_{\Omega} \; \sqrt{} \; 9 \; \infty \; \overset{\leq}{} \; \Sigma \; 2$$

MATH BREAK

Bending by Degrees

Suppose a plant has a positive phototropism and bends toward the light at a rate of 0.3 degrees per minute. How many hours will it take the plant to bend 90 degrees?

Plant Responses to the Environment

What happens when you get really cold? Do your teeth chatter as you shiver uncontrollably? If so, your brain is responding to the stimulus of cold by causing your muscles to twitch rapidly and generate warmth. Anything that causes a reaction in an organ or tissue is a stimulus. Do plant tissues respond to stimuli? They sure do! Examples of stimuli to which plants respond include light, gravity, and changing seasons.

Plant Tropisms

Some plants respond to an environmental stimulus, such as light or gravity, by growing in a particular direction. Growth in response to a stimulus is called a **tropism.** Tropisms are either positive or negative, depending on whether the plant grows toward or away from the stimulus. Plant growth toward a stimulus is a positive tropism. Plant growth away from a stimulus is a negative tropism. Two examples of tropisms are phototropism and gravitropism.

Sensing Light If you place a houseplant so that it gets light from only one direction, such as from a window, the shoot tips bend toward the light. A change in the growth of a plant that is caused by light is called **phototropism** (foh TAH troh PIZ uhm). As shown in **Figure 9,** the bending occurs when cells on one side of the shoot grow longer than cells on the other side of the shoot.

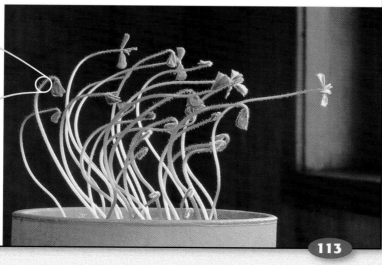

Figure 9 *The plant cells on the dark side of the shoot grow longer than the cells on the other side. This causes the shoot to bend toward the light.*

113

IS THAT A FACT!

The tallest tree ever measured was found in Watts River, in Victoria, Australia. The tree—an Australian eucalyptus *(Eucalyptus regnans)*—was 132 m tall in 1872.

Answer to MATHBREAK

5 hours

Focus

Plant Responses to the Environment

This section describes how plants may respond to light and gravity and how some plants flower in response to night length. Students will be able to describe how some plants are adapted to survive in cold weather.

🔊 Bellringer

Have students write in their ScienceLog a hypothesis explaining why some leaves of some trees undergo a color change in autumn. Have volunteers read their hypothesis aloud, and let the students know they'll be learning more about such plant responses in this section.

1) Motivate

DISCUSSION

Plant Responses Bring to class a sensitive plant, such as *Mimosa pudica.* Show students how the leaves of the plant "fold up" when they are touched. You also can demonstrate plant movement with a Venus' flytrap *(Dionaea muscipula).* Have a student volunteer gently probe inside the open "trap" with a pencil, and observe how the plant responds.

Directed Reading Worksheet Section 3

USING THE FIGURE

Have students look closely at **Figure 10.** Point out that after a few days the leaves of the plant grow upward and the roots grow downward. Explain that the plant growth is in response (positive or negative) to gravity. Ask students what other stimuli the stems might be growing in response to. (Students should reason that the stems grow upward in order for the leaves to reach sunlight and roots grow downward to reach water or moisture and to anchor the plant.)
Sheltered English

QuickLab

MATERIALS

FOR EACH GROUP:
• several potted plants
• duct tape
• cardboard

Answers to QuickLab

Observations may vary, but students should see that plant stems always try to grow perpendicular to the ground.

Gravity and light might have influenced the growth by causing the plant to grow upward.

Gravitropism benefits a plant because it directs the plant to grow so that the stems are aboveground and the roots grow downward.

---

internetconnect

SC**LINKS**
NSTA

TOPIC: Plant Tropisms
GO TO: www.scilinks.org
*sci***LINKS NUMBER:** HSTL315

TOPIC: Plant Growth
GO TO: www.scilinks.org
*sci***LINKS NUMBER:** HSTL320

Which Way Is Up? When the growth of a plant changes direction in response to the direction of gravity, the change is called **gravitropism** (GRAV i TROH PIZ uhm). The effect of gravitropism is demonstrated by the plants in **Figure 10.** A few days after a plant is placed on its side or turned upside down, the roots and shoots show a change in their direction of growth. Most shoot tips have negative gravitropism—they grow upward, away from the center of the Earth. In contrast, most root tips have positive gravitropism—they grow downward, toward the center of the Earth.

Figure 10 *Gravity is a stimulus that causes plants to change their direction of growth.*

▲ To grow away from the pull of gravity, this plant has grown upward.

This plant has recently ▶ been upside down.

QuickLab

Which Way Is Up?

Will a potted plant grow sideways? You will need **several potted plants** to find out. Use **duct tape** to secure **cardboard** around the base of each plant so that the soil will not fall out. Turn the plants on their sides and observe what happens over the next few days. Describe two stimuli that might have influenced the direction of growth. How might gravitropism benefit a plant?

✓Self-Check

1. Use the following terms to create a concept map: tropism, stimuli, light, gravity, phototropism, and gravitropism.

2. Imagine a plant in which light causes a negative phototropism. Will the plant bend to the left or to the right when light is shining only on the plant's right side?

(See page 168 to check your answers.)

Answers to Self-Check

1.

Tropism
is a plant's response to
stimuli
such as
gravity   light
which causes   which causes
gravitropism   phototropism

2. During negative phototropism, the plant would grow away from the stimulus (light), so it would be bending to the left.

Seasonal Responses

What would happen if a plant living in an area that has severe winters flowered in December? Would the plant be able to successfully produce seeds and fruits? If your answer is no, you're correct. If the plant produced any flowers at all, the flowers would probably freeze and die before they had the chance to produce mature seeds. Plants living in regions with cold winters can detect the change in seasons. How do plants do this?

As Different as Night and Day Think about what happens as the seasons change. For example, what happens to the length of the days and the nights? As autumn and winter approach, the days get shorter and the nights get longer. The opposite happens when spring and summer approach.

The difference between day length and night length is an important environmental stimulus for many plants. This stimulus can cause plants to begin reproducing. Some plants flower only in late summer or early autumn, when the night length is long. These plants are called short-day plants. Examples of short-day plants include poinsettias (shown in **Figure 11**), ragweed, and chrysanthemums. Other plants flower in spring or early summer, when night length is short. These plants are called long-day plants. Clover, spinach, and lettuce are examples of long-day plants.

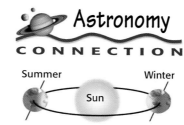

Astronomy CONNECTION

Summer   Winter

Sun

The seasons are caused by Earth's tilt and its orbit around the sun. We have summer when the Northern Hemisphere is tilted toward the sun and the sun's energy falls more directly on the Northern Hemisphere. While the Northern Hemisphere experiences the warm season, the Southern Hemisphere experiences the cold season. The opposite of this occurs when the Northern Hemisphere is tilted away from the sun.

Figure 11 *Night length determines when poinsettias flower.*

Early summer

◀ In the early summer, night length is short. At this time, poinsettia leaves are all green, and there are no flowers.

Late fall

◀ Poinsettias flower in the fall, when nights are longer. The leaves surrounding the flower clusters turn red. Professional growers use artificial lighting to control the timing of the color change.

115

IS THAT A FACT!

Short-day plants are really "long-night" plants. The number of hours of darkness is actually more important to the plant than the number of hours of daylight. The poinsettias shown in **Figure 11,** for example, will not flower until the amount of darkness is 14 hours or longer. If that period of darkness is interrupted for any reason—for even a brief time—the plant will not flower.

REAL-WORLD CONNECTION

Many greenhouses and nurseries cause plants to bloom at times when they would not normally bloom. By controlling the temperature and the amount of light plants receive, growers can force poinsettias, tulips, daffodils, and other plants to bloom. Have students contact a local greenhouse that grows poinsettias or lilies. Have them find out what factors the growers control to force the plants to bloom.

Homework

Investigate Plant Growth
Have students research how the lifespan of a plant is measured in one-year growing seasons. Ask students to describe the life cycle of plants in each category and to provide a few examples of each.

annuals (plants that complete their entire life cycle within one growing season—marigolds, snapdragons, sunflowers)

biennials (plants that complete their life cycle in two growing seasons—hollyhock)

perennials (plants that repeat their life cycle every year—roses, daisies, irises)

Encourage students to include drawings that illustrate when each plant begins to grow, produces seeds, and dies.

Teaching Transparency 53 "Night Length and Blooming Flowers"

Interactive Explorations CD-ROM "How's It Growing?"

Answer to APPLY

Sample answers include the following: experiments in which entire trees were covered for part of the day to prevent them from getting sunlight; experiments in which entire trees were exposed to artificial sunlight to determine if trees could be kept from losing their leaves; experiments in which some of the leaves were protected or given artificial sunlight to determine if some leaves could be kept on the tree, even when others fell off.

READING 📖 STRATEGY

Prediction Guide Before students read this page, ask the following questions:

- What kind of tree is an evergreen? Give two examples. (Evergreens are trees that have leaves adapted to survive throughout the year. Examples include conifers, such as pines and firs.)

- What are deciduous trees? (Deciduous trees lose all their leaves at the same time each year.)

MISCONCEPTION ALERT

The colored part of the poinsettia is not the flower, as most people think it is. The colored parts are really types of leaves called bracts. The flowers are the tiny berrylike structures at the center of the colored bracts.

Teaching Transparency 53 "The Change of Seasons and Pigment Color"

Reinforcement Worksheet "How Plants Respond to Change"

Can Trees Tell Time?

One fall afternoon, Holly looks into her backyard. She notices that a tree that was full of leaves the week before is now completely bare. What caused the tree to drop all its leaves?

Holly came up with the following hypothesis:

Each leaf on the tree was able to sense day length. When the day length became short enough, each leaf responded by falling.

Design an experiment that would test Holly's hypothesis.

Seasonal Changes in Leaves All trees lose their leaves at some time. Some trees, such as pine and holly, shed some of their leaves year-round so that some leaves are always present on the tree. These trees are called **evergreen.** Evergreen trees have leaves adapted to survive throughout the year.

Other trees, such as the maple tree in **Figure 12,** are **deciduous** and lose all their leaves at the same time each year. Deciduous trees usually lose their leaves before winter begins. In tropical climates that have wet and dry seasons, deciduous trees lose their leaves before the dry season. Having bare branches during the winter or during the dry season may reduce the water lost by transpiration. The loss of leaves helps plants survive low temperatures or long periods without rain.

Figure 12 *The leaves of some deciduous trees, like the maple shown here, change from green to orange in autumn. In winter, the maple is bare.*

IS THAT A FACT!

When leaves change color in the fall, they make a beautiful display. Here are the autumn leaf colors of some common trees.

- flame red and orange—sugar maple, sumac
- dark red—dogwood, red maple, scarlet oak
- yellow—poplar, willow, birch
- plum purple—ash
- tan or brown—oak, beech, elm, hickory

The leaves of some trees, such as locusts, stay green until the leaves drop. The leaves of black walnut trees fall before they change color.

As shown in **Figure 13,** leaves often change color before they fall. As autumn approaches, chlorophyll, the green pigment used in photosynthesis, breaks down. As chlorophyll is lost from leaves, other yellow and orange pigments are revealed. These pigments were always present in the leaves but were hidden by the green chlorophyll. Some leaves also have red pigments, which also become visible when chlorophyll is broken down.

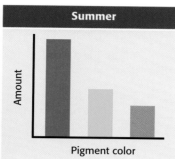
Summer

Amount
Pigment color

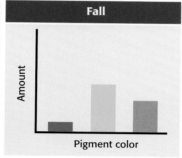
Fall

Amount
Pigment color

BRAIN FOOD

In autumn, trees growing near streetlights keep their leaves longer than their rural counterparts.

Figure 13 *The breakdown of chlorophyll in the autumn is a seasonal response in many trees. As the amount of chlorophyll in leaves decreases, other pigments become visible.*

SECTION REVIEW

1. What are the effects of the tropisms caused by light and gravity?

2. What is the difference between a short-day plant and a long-day plant?

3. How does the loss of leaves help a plant survive winter or long periods without rain?

4. **Applying Concepts** If a plant does not flower when exposed to 12 hours of daylight but does flower when exposed to 15 hours of daylight, is it a short-day plant or a long-day plant?

internetconnect

*sci*LINKS.
NSTA

TOPIC: Plant Tropisms, Plant Growth
GO TO: www.scilinks.org
*sci*LINKS NUMBER: HSTL315, HSTL320

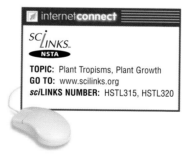

▼ *Answers to Section Review*

1. Phototropism is caused by light. The cells on the dark side of the stem grow longer than the cells on the side facing the light, causing the stem to bend toward the light. Gravitropism is caused by gravity. Parts of the plant growing above the surface grow away from the pull of gravity. Parts below the surface grow toward it.

2. Plants that flower in only late summer or early autumn, when the night length is long, are called short-day plants. Plants that flower in the spring or early summer when night length is short, are called long-day plants.

3. Dropping its leaves prevents the tree from losing water through transpiration.

4. It is a long-day plant.

3) Extend

GOING FURTHER

Writing Have students explore other kinds of seasonal changes exhibited in plants. For example, encourage students to find out what conditions prompt bulbs to emerge in the spring and what conditions prompt seeds of flowering plants to germinate.

4) Close

Quiz

1. Why do some leaves change color in the fall? (Green chlorophyll in the leaves begins to break down, and other pigments become visible.)

2. What is the difference between an evergreen tree and a deciduous tree? (An evergreen tree has leaves adapted to survive throughout the year; they shed some of their leaves year-round. A deciduous tree loses all its leaves at the same time each year.)

3. What is a tropism in a plant? (A tropism is either a positive or negative response to an environmental stimulus, such as light or gravity.)

ALTERNATIVE ASSESSMENT

Have students make labeled drawings in their ScienceLog that illustrate plants showing positive and negative phototropism and positive and negative gravitropism.
Sheltered English

Critical Thinking Worksheet
"Space Plants"

Section 3 • Plant Responses to the Environment **117**

Skill Builder Lab

Food Factory Waste

Food Factory Waste
Teacher's Notes

Time Required

One 45-minute class period and about 15 minutes per day for 5 days

Lab Ratings

EASY ———————————→ HARD

TEACHER PREP
STUDENT SET-UP
CONCEPT LEVEL
CLEAN UP

MATERIALS

The materials listed on the student page are enough for 1 student.

Elodea is a common aquarium plant and can be found at some pet stores and most places that sell aquarium fish.

A 5 percent solution of baking soda and water can be made by dissolving 5 g of baking soda in 95 mL of water.

Safety Caution

Remind students to review all safety cautions and icons before beginning this lab activity.

Food Factory Waste

Plants use photosynthesis to produce food. We cannot live without the waste products from this process. In this activity, you will observe the process of photosynthesis and determine the rate of photosynthesis for *Elodea*.

MATERIALS

- 500 mL of 5% baking soda-and-water solution
- 600 mL beaker
- 20 cm long *Elodea* sprigs (2–3)
- glass funnel
- test tube
- metric ruler

Procedure

1. Add 450 mL of baking soda–and–water solution to a beaker.

2. Put two or three sprigs of *Elodea* in the beaker. The baking soda will provide the *Elodea* with the carbon dioxide it needs for photosynthesis.

3. Place the wide end of the funnel over the *Elodea*. The end of the funnel with the small opening should be pointing up. The *Elodea* and the funnel should be completely covered by the solution.

4. Fill a test tube with the remaining baking soda–and–water solution. Place your thumb over the end of the test tube. Turn the test tube upside down, taking care that no air enters. Hold the opening of the test tube under the solution and place the test tube over the small end of the funnel. Try not to let any solution out of the test tube as you do this.

5. Place the beaker setup in a well-lit area near a lamp or in direct sunlight.

6. Prepare a data table similar to the one below.

Amount of Gas Present in the Test Tube		
Days of exposure to light	Total amount of gas present (mm)	Amount of gas produced per day (mm)
0		
1		
2		
3		
4		
5		

DO NOT WRITE IN BOOK

Datasheets for LabBook

Science Skills Worksheet
"Introduction to Graphs"

CLASSROOM TESTED & APPROVED

David Sparks
Redwater Junior High School
Redwater, Texas

7. Record that there was 0 mm of gas in the test tube on day 0. (If you were unable to place the test tube without getting air in the tube, measure the height of the column of air in the test tube in millimeters. Record this value for day 0.) Measure the gas in the test tube from the middle of the curve on the bottom of the upside-down test tube to the level of the solution.

8. For days 1 through 5, measure the amount of gas in the test tube. Record the measurements in your data table under the heading "Total amount of gas present (mm)."

9. Calculate the amount of gas produced each day by subtracting the amount of gas present on the previous day from the amount of gas present today. Record these amounts under the heading "Amount of gas produced per day (mm)."

10. Plot the data from your table on a graph.

Analysis

11. Using information from your graph, describe what happened to the amount of gas in the test tube.

12. How much gas was produced in the test tube after day 5?

13. Write the equation for photosynthesis. Explain each part of the equation. For example, what "ingredients" are necessary for photosynthesis to take place? What substances are produced by photosynthesis? What gas is produced that we need in order to live?

14. Write a report describing your experiment, your results, and your conclusions.

Going Further

Hydroponics is the growing of plants in nutrient-rich water. Research hydroponic techniques, and try to grow a plant without soil.

Answers

11. Students' graphs should show a gradual increase in the amount of gas in the test tube.

12. Answers will vary according to variables in the classroom, such as the amount of light and temperature.

13. $6CO_2 + 6H_2O + \text{light energy} \rightarrow C_6H_{12}O_6 + 6O_2$

    CO_2 is carbon dioxide and comes from the baking soda solution. H_2O is the water in the solution. Light energy comes from the sun. $C_6H_{12}O_6$ is sugar (glucose), and O_2 is oxygen. Photosynthesis produces sugar and oxygen. Plants produce oxygen as a byproduct of photosynthesis; oxygen is the gas that we breathe and cannot live without.

119

Lab Notes

You may need to have students practice placing the test tube over the inverted funnel. It may take two or three tries to get the test tube over the funnel stem without letting any air into the tube. First fill the test tube with the solution. Place your thumb over the opening tightly so no air can get in. Submerge your thumb and the top of the test tube underwater. Once the top of the test tube is underwater you can release your thumb to maneuver the test tube over the stem of the funnel. Be sure you have the *Elodea* in place under the funnel before you begin!

You may wish to write the equation for photosynthesis on the board.

Chapter Highlights

Chapter Highlights

VOCABULARY DEFINITIONS

SECTION 1

dormant describes an inactive state of a seed

SECTION 2

chlorophyll a green pigment in chloroplasts that absorbs light energy for photosynthesis

cellular respiration the process of producing ATP in the cell from oxygen and glucose; releases carbon dioxide and water

stomata openings in the epidermis of a leaf that allow carbon dioxide to enter the leaf and allow water and oxygen to leave the leaf

transpiration the loss of water from plant leaves through openings called stomata

---

SECTION 1

Vocabulary

dormant (p. 108)

Section Notes

- Sexual reproduction in flowering plants requires pollination and fertilization. Fertilization is the joining of an egg and a sperm cell.

- After fertilization has occurred, the ovules develop into seeds that contain plant embryos. The ovary develops into a fruit.

- Once seeds mature, they may become dormant. Seeds sprout when they are in an environment with the proper temperature and the proper amounts of water and oxygen.

- Many flowering plants can reproduce asexually without flowers.

SECTION 2

Vocabulary

chlorophyll (p. 110)
cellular respiration (p. 111)
stomata (p. 112)
transpiration (p. 112)

Section Notes

- During photosynthesis, leaves absorb sunlight and form glucose from carbon dioxide and water.

- During cellular respiration, a plant uses oxygen and releases carbon dioxide and water. Glucose is converted into a form of energy that cells can use.

- Plants take in carbon dioxide and release oxygen and water through stomata in their leaves.

Labs

Weepy Weeds (p. 138)

☑ Skills Check

Visual Understanding

CIRCLE GRAPH A circle graph is a great visual for illustrating fractions without using numbers. Each circle graph on page 115 represents a 24-hour period. The blue area represents the fraction of time in which there is no sunlight, and the gold area represents the fraction of time in which there is light. As shown by the graph, early summer is about two-thirds day and one-third night.

BAR GRAPHS As shown on page 117, bar graphs are often used to compare numbers. The graph at right compares the success rate of flower seeds from five seed producers. As shown by the bar height, company D, at about 88 percent, had the highest rate of success.

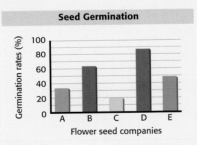

Seed Germination

Lab and Activity Highlights

Food Factory Waste   PG 118

Weepy Weeds   PG 138

 Datasheets for LabBook
(blackline masters for these labs)

SECTION 3

Vocabulary

tropism *(p. 113)*

phototropism *(p. 113)*

gravitropism *(p. 114)*

evergreen *(p. 116)*

deciduous *(p. 116)*

Section Notes

- A tropism is plant growth in response to an environmental stimulus, such as light or gravity. Plant growth toward a stimulus is a positive tropism. Plant growth away from a stimulus is a negative tropism.

- Phototropism is growth in response to the direction of light. Gravitropism is growth in response to the direction of gravity.

- The change in the amount of daylight and darkness that occurs with changing seasons often controls plant reproduction.

- Evergreen plants have leaves adapted to survive throughout the year. Deciduous plants lose their leaves before cold or dry seasons. The loss of leaves helps deciduous plants survive low temperatures and dry periods.

VOCABULARY DEFINITIONS, *continued*

SECTION 3

tropism a change in the growth of a plant in response to a stimulus

phototropism a change in the growth of a plant in response to light

gravitropism a change in the growth of a plant in response to gravity

evergreen describes trees that keep their leaves year-round

deciduous describes trees that have leaves that change color in autumn and fall off in winter

Vocabulary Review Worksheet

Blackline masters of these Chapter Highlights can be found in the **Study Guide.**

---

internet**connect**

go.hrw.com

GO TO: go.hrw.com

Visit the **HRW** Web site for a variety of learning tools related to this chapter. Just type in the keyword:

KEYWORD: HSTPL2

SCI LINKS

N S T A

GO TO: www.scilinks.org

Visit the **National Science Teachers Association** on-line Web site for Internet resources related to this chapter. Just type in the *sci*LINKS number for more information about the topic:

TOPIC:	*sci*LINKS NUMBER:
Reproduction of Plants	HSTL305
Photosynthesis	HSTL310
Plant Tropisms	HSTL315
Plant Growth	HSTL320

121

Lab and Activity Highlights

LabBank

Labs You Can Eat, Not Just Another Nut

Ecolabs & Field Activities,
Recycle! Make Your Own Paper

Long-Term Projects & Research Ideas,
Plant Partners

Interactive Explorations CD-ROM

CD 2, Exploration 8, "How's It Growing?"

Chapter Review
Answers

USING VOCABULARY

1. dormant
2. photosynthesis
3. transpiration
4. phototropism
5. evergreen

UNDERSTANDING CONCEPTS

Multiple Choice

6. a
7. d
8. b
9. d
10. c
11. d

Short Answer

12. The cuticle is the waxy covering that prevents the leaf from losing moisture. The stomata are openings in the epidermis of the leaf that open and close, allowing only a certain amount of moisture to escape the leaf. Transpiration is the process that occurs when moisture escapes through the stomata.

13. The stimulus is light. The plant hormone auxin moves away from the light side to the dark side of the plant, causing the cells on the dark side to grow longer.

14. Phototropism is a positive tropism, and gravitropism in stems is a negative tropism.

Chapter Review

USING VOCABULARY

To complete the following sentences, choose the correct term from each pair of terms listed below:

1. After seeds develop fully, and before they sprout, they may become ___?___. (*deciduous* or *dormant*)

2. During ___?___, energy from sunlight is used to make sugar. (*photosynthesis* or *phototropism*)

3. The loss of water through stomata is called ___?___. (*transpiration* or *tropism*)

4. A change in plant growth in response to the direction of light is called ___?___. (*gravitropism* or *phototropism*)

5. Plants that have leaves year-round are ___?___. (*deciduous* or *evergreen*)

UNDERSTANDING CONCEPTS

Multiple Choice

6. The cells that open and close the stomata are the
   a. guard cells.
   b. xylem cells.
   c. cuticle cells.
   d. mesophyll cells.

7. Plant cells need carbon dioxide, which is used for
   a. cellular respiration.　c. fertilization.
   b. phototropism.　　　　d. photosynthesis.

8. When chlorophyll breaks down,
   a. pollination occurs.
   b. other pigments become visible.
   c. red pigments disappear.
   d. photosynthesis occurs.

9. Which of the following sequences shows the correct order of events that occur after an insect brings pollen to a flower?
   a. germination, fertilization, pollination
   b. fertilization, germination, pollination
   c. pollination, germination, fertilization
   d. pollination, fertilization, germination

10. When the amount of water transpired from a plant's leaves is greater than the amount absorbed by its roots,
   a. the cuticle conserves water.
   b. the stem exhibits positive gravitropism.
   c. the plant wilts.
   d. the plant recovers from wilting.

11. Ovules develop into
   a. fruits.　　　　c. flowers.
   b. ovaries.　　　d. seeds.

Short Answer

12. What is the relationship between transpiration, the cuticle, and the stomata?

13. What is the stimulus in phototropism? What is the plant's response to the stimulus?

14. Give an example of a positive tropism and a negative tropism.

122

Concept Mapping

15. Use the following terms to create a concept map: plantlets, flower, seeds, ovules, plant reproduction, asexual, runners.

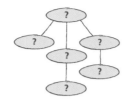

CRITICAL THINKING AND PROBLEM SOLVING

Write one or two sentences to answer the following questions:

16. Many plants that live in regions that experience severe winters have seeds that will not germinate at any temperature unless the seeds have been exposed first to a long period of cold. How might this characteristic help new plants survive?

17. If you wanted to make poinsettias bloom and turn red in the summer, what would you have to do?

18. What benefit is there for a plant's shoots to have positive phototropism? What benefit is there for its roots to have positive gravitropism?

MATH IN SCIENCE

19. If a particular leaf has a surface area of 8 cm², what is its surface area in square millimeters? (Hint: 1 cm² = 100 mm².)

20. Leaves have an average of 100 stomata per square millimeter of surface area. How many stomata would you expect the leaf in question 19 to possess?

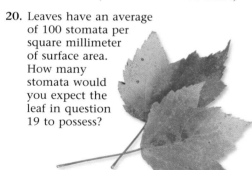

INTERPRETING GRAPHICS

Look at the illustrations below, and then answer the questions that follow. The illustration shows part of an experiment on phototropism in young plants. In part (1), the young plants have just been placed in the light after being in the dark. The shoot tip of one plant is cut off. The other tip is not cut. In part (2), the plants from part (1) are exposed to light from one direction.

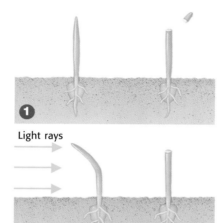

Light rays

21. Why did the plant with the intact tip bend toward the light?

22. Why did the plant with the removed tip remain straight?

Reading Check-up

Take a minute to review your answers to the Pre-Reading Questions found at the bottom of page 104. Have your answers changed? If necessary, revise your answers based on what you have learned since you began this chapter.

Concept Mapping

15. An answer to this exercise can be found at the front of this book.

CRITICAL THINKING AND PROBLEM SOLVING

16. Exposure to a long period of cold may prevent a seed from sprouting before winter. If a seed sprouted in the fall, it might not have time to produce more seeds before it dies.

17. You would have to use artificial means to provide a long period of darkness and a short period of light.

18. Positive phototropism helps the leaves get exposure to sunlight so that photosynthesis can occur. Gravitropism helps the roots grow in the correct direction so that they will have access to water and nutrients in the soil.

MATH IN SCIENCE

19. 800 mm²

20. 80,000 stomata

INTERPRETING GRAPHICS

21. The plant hormone auxin caused the shoot to bend toward the light.

22. The plant hormone auxin is produced in the shoot tip. If the tip is removed, the shoot will not bend toward the light.

Concept Mapping Transparency 13

Blackline masters of this Chapter Review can be found in the **Study Guide.**

Teaching Strategy

To help your students understand how the parts of Meyerowitz's mustard flowers are all in the wrong places, you may wish to review with your students the appropriate locations of the parts of a flower. The parts of a flower are modified leaves, usually found in four successive whorls or rings. Create a diagram that shows four concentric circles. Label the innermost circle "Pistil." Label the next circle "Stamens." (Stamens produce pollen, and each stamen has two parts, the anther and the filament.) Label the third circle "Petals." (Petals are leaflike structures that surround the carpels and stamens. Their colors and shapes give many flowers a distinctive appearance.) Label the outermost circle "Sepals." (The sepals protect the flower when it is a bud.) Explain that the inner two circles mark the locations of the reproductive organs of the flower and that the petals and the sepals are called nonessential flower parts because they are not directly involved in reproduction.

WEIRD SCIENCE

MUTANT MUSTARD

The tiny mustard flowers grown by Elliot Meyerowitz are horribly deformed. You may think they are the result of a terrible accident, but Meyerowitz created these mutants on purpose. In fact, he is very proud of these flowers because they may help him solve an important biological mystery.

▲ *Elliot Meyerowitz, shown here in his laboratory, has raised about a million individual specimens of a mustard variety known as Arabidopsis thaliana.*

Normal and Abnormal Flowers

Normally, mustard flowers have four distinct parts that are arranged in a specific way. Many of the plants grown by Meyerowitz and his colleagues, however, are far from normal. Some have leaves in the center of their flowers. Others have seed-producing ovaries where the petals should be. At first glance, the arrangement of the parts seems random, but the structure of each flower has actually been determined by a small number of genes.

A Simple Model

After many years of careful studies, Meyerowitz and his colleagues have identified most of the genes that control the mustard flower's development. With this information, Meyerowitz has discovered patterns that have led to a surprisingly simple model. The model

points to just three classes of genes that determine what happens to the various parts of a flower as it develops. He learned that if one or more of those gene classes is inactivated, a mutant mustard plant results.

Pieces of an Old Puzzle

By understanding how genes shape the growth of flowers, Meyerowitz hopes to add pieces to a long-standing puzzle involving the origin of flowering plants. Scientists estimate that flowering plants first appeared on Earth about 125 million years ago and that they quickly spread to become the dominant plants on Earth. By studying which genes produce flowers in present-day plants, Meyerowitz and his colleagues hope to learn how flowering plants evolved in the first place.

Meyerowitz's mutant plants are well qualified to add to our understanding of plant genetics. But don't look for these mustard plants in your local flower shop. These strange mutants won't win any prizes for beauty!

▲ *Meyerowitz alters the genes of a mustard plant so that it develops a mutant flower. The inset shows a normal flower.*

Think About It

▶ It is possible to genetically change a plant. What are some possible risks of such a practice?

Answer to Think About It

Accept all reasonable responses. Some students may point out that the process of growing many generations of plants may be quite lengthy. In addition, it often takes many unsuccessful tests to develop a successful hypothesis. Students may also point out that genetically altering a plant may have positive or negative repercussions in terms of the usefulness of the plant.

EYE ON THE ENVIRONMENT

A Rainbow of Cotton

Think about your favorite T-shirt. Chances are, it's made of cotton and brightly colored. The fibers in cotton plants, however, are naturally white. They must be dyed with chemicals—often toxic ones—to create the bright colors seen in T-shirts and other fabrics. To minimize the use of toxic chemicals, an ingenious woman named Sally Fox had an idea: What if you could grow the cotton *already colored*?

Learning from the Past

Cotton fibers come from the plant's seed pods, or *bolls.* Bolls are a little bigger than a golf ball and open at maturity to reveal a fuzzy mass of fibers and seeds. Once the seeds are removed, the fibers can be twisted into yarn and used to make many kinds of fabric. Sally Fox began her career as an *entomologist,* a scientist who studies insects. She first found out about colored cotton while studying pest resistance. Although most of the cotton grown for textiles is white, different shades of cotton have been harvested by Native Americans for centuries. These types of cotton showed some resistance to pests but had fibers too short to be used by the textile industry.

In 1982, Fox began the very slow process of crossbreeding different varieties of cotton to produce strains that were both colored *and* long-fibered. Her cotton is registered under the name FoxFibre® and has earned her high praise.

Solutions to Environmental Problems

The textile industry is the source of two major environmental hazards. The first hazard is the dyes used for cotton fabrics. The second is the pesticides that are required for growing cotton. These pesticides, like the dyes, can cause damage to both living things and natural resources, such as water and land.

Fox's cotton represents a solution to both of these problems. First, since the cotton is naturally colored, no dyes are necessary. Second, the native strains of cotton from which she bred her plants passed on their natural pest resistance. Thus, fewer pesticides are necessary to grow her cotton successfully.

Sally Fox's efforts demonstrate that with ingenuity and patience, science and agriculture can work together in new ways to offer solutions to environmental problems.

▲ *Sally Fox in a field of colored cotton*

Some Detective Work

▶ Like Fox's cotton, many types of plants and breeds of domesticated animals have been created through artificial selection. Research to find out where and when your favorite fruit or breed of dog was first established.

Answers to Some Detective Work

One example is the dachshund. It was initially bred in Germany in the 1700s, where it was used to hunt badgers. Its tubular body and short legs are ideal for following prey into burrows. The word *dachshund* is German for "badger hound."

Background

Cotton has many properties that make it a superior material for fabrics. Cotton fabrics are durable, washable, and comfortable. Cotton clothing keeps you cool in the summer because it "breathes" well and allows the moisture to evaporate quickly. Some other uses for cotton include fine yarns, carpets, and blends with other fabrics.

Sally Fox grows cotton in Arizona and New Mexico. Fox is continuing to refine her crop by trying to breed fire-resistant varieties as well as new colors.

Teaching Strategy

Give students samples of different fabrics to touch and observe. Fabrics could include cotton, linen, polyester, nylon, silk, leather, or suede. Then have students classify the fabrics based on their origin—plant, animal, or synthetic. Ask students the following question:

What are the advantages of fabrics made from plants? What are the disadvantages? (Students should recognize that plant-based fabrics have different texture, come from a renewable resource, and are cheaper than animal-based fabrics. They should also recognize that plant-based fabrics may be less durable and more prone to stains than other fabrics.)

SAFETY FIRST!

Exploring, inventing, and investigating are essential to the study of science. However, these activities can also be dangerous. To make sure that your experiments and explorations are safe, you must be aware of a variety of safety guidelines.

You have probably heard of the saying, "It is better to be safe than sorry." This is particularly true in a science classroom where experiments and explorations are being performed. Being uninformed and careless can result in serious injuries. Don't take chances with your own safety or with anyone else's.

Following are important guidelines for staying safe in the science classroom. Your teacher may also have safety guidelines and tips that are specific to your classroom and laboratory. Take the time to be safe.

Safety Rules!

Start Out Right

Always get your teacher's permission before attempting any laboratory exploration. Read the procedures carefully, and pay particular attention to safety information and caution statements. If you are unsure about what a safety symbol means, look it up or ask your teacher. You cannot be too careful when it comes to safety. If an accident does occur, inform your teacher immediately, regardless of how minor you think the accident is.

Safety Symbols

All of the experiments and investigations in this book and their related worksheets include important safety symbols to alert you to particular safety concerns. Become familiar with these symbols so that when you see them, you will know what they mean and what to do. It is important that you read this entire safety section to learn about specific dangers in the laboratory.

If you are instructed to note the odor of a substance, wave the fumes toward your nose with your hand. Never put your nose close to the source.

Eye protection	Clothing protection	Hand safety
Heating safety	Electric safety	Chemical safety
Animal safety	Sharp object	Plant safety

Eye Safety

Wear safety goggles when working around chemicals, acids, bases, or any type of flame or heating device. Wear safety goggles any time there is even the slightest chance that harm could come to your eyes. If any substance gets into your eyes, notify your teacher immediately, and flush your eyes with running water for at least 15 minutes. Treat any unknown chemical as if it were a dangerous chemical. Never look directly into the sun. Doing so could cause permanent blindness.

Avoid wearing contact lenses in a laboratory situation. Even if you are wearing safety goggles, chemicals can get between the contact lenses and your eyes. If your doctor requires that you wear contact lenses instead of glasses, wear eye-cup safety goggles in the lab.

Safety Equipment

Know the locations of the nearest fire alarms and any other safety equipment, such as fire blankets and eyewash fountains, as identified by your teacher, and know the procedures for using them.

Be extra careful when using any glassware. When adding a heavy object to a graduated cylinder, tilt the cylinder so the object slides slowly to the bottom.

Neatness

Keep your work area free of all unnecessary books and papers. Tie back long hair, and secure loose sleeves or other loose articles of clothing, such as ties and bows. Remove dangling jewelry. Don't wear open-toed shoes or sandals in the laboratory. Never eat, drink, or apply cosmetics in a laboratory setting. Food, drink, and cosmetics can easily become contaminated with dangerous materials.

Certain hair products (such as aerosol hair spray) are flammable and should not be worn while working near an open flame. Avoid wearing hair spray or hair gel on lab days.

Sharp/Pointed Objects

Use knives and other sharp instruments with extreme care. Never cut objects while holding them in your hands. Place objects on a suitable work surface for cutting.

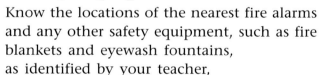

Heat

Wear safety goggles when using a heating device or a flame. Whenever possible, use an electric hot plate as a heat source instead of an open flame. When heating materials in a test tube, always angle the test tube away from yourself and others. In order to avoid burns, wear heat-resistant gloves whenever instructed to do so.

Electricity

Be careful with electrical cords. When using a microscope with a lamp, do not place the cord where it could trip someone. Do not let cords hang over a table edge in a way that could cause equipment to fall if the cord is accidentally pulled. Do not use equipment with damaged cords. Be sure your hands are dry and that the electrical equipment is in the "off" position before plugging it in. Turn off and unplug electrical equipment when you are finished.

Chemicals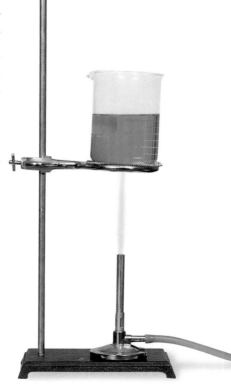

Wear safety goggles when handling any potentially dangerous chemicals, acids, or bases. If a chemical is unknown, handle it as you would a dangerous chemical. Wear an apron and safety gloves when working with acids or bases or whenever you are told to do so. If a spill gets on your skin or clothing, rinse it off immediately with water for at least 5 minutes while calling to your teacher.

Never mix chemicals unless your teacher tells you to do so. Never taste, touch, or smell chemicals unless you are specifically directed to do so. Before working with a flammable liquid or gas, check for the presence of any source of flame, spark, or heat.

Animal Safety

Always obtain your teacher's permission before bringing any animal into the school building. Handle animals only as your teacher directs. Always treat animals carefully and with respect. Wash your hands thoroughly after handling any animal.

Plant Safety

Do not eat any part of a plant or plant seed used in the laboratory. Wash hands thoroughly after handling any part of a plant. When in nature, do not pick any wild plants unless your teacher instructs you to do so.

Glassware

Examine all glassware before use. Be sure that glassware is clean and free of chips and cracks. Report damaged glassware to your teacher. Glass containers used for heating should be made of heat-resistant glass.

Does It All Add Up?
Teacher's Notes

Time Required

One 45-minute class period

Lab Ratings

EASY ——————————→ HARD

TEACHER PREP 🍶🍶
STUDENT SET-UP 🍶
CONCEPT LEVEL 🍶🍶
CLEAN UP 🍶

MATERIALS

The materials listed on the student page are enough for 1 student or 1 group of students. Liquid A is plain water. Liquid B is either isopropyl alcohol or denatured ethyl alcohol. Safety thermometers are recommended for this lab.

Safety Caution

Remind students to review all safety cautions and icons before beginning this lab activity.

Caution students to handle mercury thermometers with care. Alcohol is flammable and poisonous. Students should wear goggles and aprons at all times during this lab. A fire extinguisher and fire blanket should be nearby. Know how to use them. The room should be well-ventilated, and students should be familiar with evacuation procedures.

SKILL BUILDER

Does It All Add Up?

Your math teacher won't tell you this, but did you know that sometimes 2 + 2 does not equal 4?! (Well, it really does, but sometimes it doesn't *appear* to equal 4.) In this experiment, you will use the scientific method to predict, measure, and observe the mixing of two unknown liquids. You will learn that a scientist does not set out to prove a hypothesis, but rather to test it, and sometimes the results just don't seem to add up!

Make Observations

1. Examine the two mystery liquids in the graduated cylinders given to you by your teacher.
   Caution: Do not taste, touch, or smell the liquids.

2. In your ScienceLog, write down as many observations as you can about each liquid. Are the liquids bubbly? What color are they? What is the exact volume of each liquid? Touch the graduated cylinders. Are they hot or cold?

3. Pour exactly 25 mL of liquid A into each of two graduated cylinders. Combine these samples in one of the graduated cylinders, and record the final volume in your ScienceLog. Repeat this step for liquid B.

Form a Hypothesis

4. Based on your observations and on prior experience, what do you expect the volume to be when you pour these two liquids together?

Materials

- 75 mL of liquid A
- 75 mL of liquid B
- 100 mL graduated cylinders (7)
- glass-labeling marker
- Celsius thermometer
- protective gloves

Make a Prediction

5. Make a prediction based on your hypothesis using an "if-then" format. Explain why you have made your prediction.

CLASSROOM TESTED & APPROVED

Kevin McCurdy
Elmwood Junior High School
Rogers, Arkansas

Test the Hypothesis

6. In your ScienceLog, make a data table similar to the one below to record your predictions and observations.

	Contents of cylinder A	Contents of cylinder B	Mixing results: predictions	Mixing results: observations
Volume				
Appearance				
Temperature				

DO NOT WRITE IN BOOK

7. Carefully pour exactly 25 mL of Liquid A into a 50 mL graduated cylinder. Mark this cylinder "A." Record its volume, appearance, and temperature in the data table.

8. Carefully pour exactly 25 mL of Liquid B into another 50 mL graduated cylinder. Mark this cylinder "B." Record its volume, appearance, and temperature in the data table.

9. Mark the empty third cylinder "A + B."

10. In the "Mixing results: predictions" column in your table, record the prediction you made earlier. Each classmate may have made a different prediction.

11. Carefully pour the contents of both cylinders into the third graduated cylinder.

12. Observe and record the total volume, appearance, and temperature in the "Mixing results: observations" column of the table.

Analyze the Results

13. Discuss your predictions as a class. How many different predictions were there? Which predictions were supported by testing? Did any of your measurements surprise you?

Draw Conclusions

14. Was your hypothesis supported? Explain why this may have happened.

15. Explain the value of incorrect predictions.

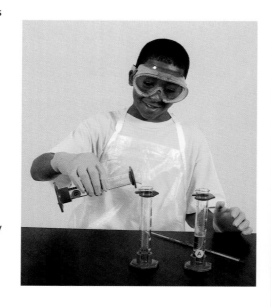

Lab Notes

This lab will be of interest because 25 mL of liquid A + 25 mL of liquid B do not make 50 mL of the mixture! Spaces between molecules of alcohol become filled with water molecules, resulting in less volume. An analogy would be mixing 25 mL of marbles with 10 mL of BBs. The BBs will settle between the marbles, and the result will be less volume. The alcohol-and-water mixture will be cloudy and bubbly for a brief time due to the sudden decrease of volume, leaving tiny bubbles.

Datasheets for LabBook
Datasheet

Science Skills Worksheet
"Working with Hypotheses"

Science Skills Worksheet
"Measuring"

131

Answers

All answers in this lab are based on student observations and will vary. Students may make some unusual predictions. You may want to lead them into questions about volume. Encourage them to think of as many ways to observe and characterize the two liquids as possible.

15. Incorrect predictions can lead to a new understanding of the way things work.

SKILL BUILDER

Graphing Data

When performing an experiment it is usually necessary to collect data. To understand the data, it is often good to organize them into a graph. Graphs can show trends and patterns that you might not notice in a table or list. In this exercise, you will practice collecting data and organizing the data into a graph.

Materials

- 200 mL of water
- 400 mL beaker
- ice
- Celsius thermometer with a clip
- hot plate
- clock or watch with a second hand
- graph paper
- heat-resistant gloves

Time Required

One 45-minute class period

Lab Ratings

EASY ——————————→ HARD

TEACHER PREP   🧪🧪
STUDENT SET-UP 🧪
CONCEPT LEVEL   🧪
CLEAN UP   🧪

Safety Caution

Remind students to review all safety cautions and icons before beginning this lab activity.

Caution students to exercise proper care when handling the beaker of hot water. Also, caution students to be careful when they are moving around an electrical cord. A clip that will hold the thermometer to the side of the beaker and off the bottom of the beaker while it is heating or cooling is safer and more accurate than a thermometer simply propped up inside the beaker. This is also good scientific practice to model.

Procedure

1. Pour 200 mL of water into a 400 mL beaker. Add ice to the beaker until the water line is at the 400 mL mark.

2. Place a Celsius thermometer into the beaker. Use a thermometer clip to prevent the thermometer from touching the bottom of the beaker. Record the temperature of the ice water in your ScienceLog.

3. Place the beaker and thermometer on a hot plate. Turn the hot plate on medium heat and record the temperature every minute until the water temperature reaches 100°C.

4. Using heat-resistant gloves, remove the beaker from the hot plate. Continue to record the temperature of the water each minute for 10 more minutes.
   Caution: Don't forget to turn off the hot plate.

5. On a piece of graph paper, create a graph similar to the one below. Label the horizontal axis (the *x*-axis) "Time (min)," and mark the axis in increments of 1 minute as shown. Label the vertical axis (the *y*-axis) "Temperature (°C)," and mark the axis in increments of ten degrees as shown.

6. Find the 1-minute mark on the *x*-axis, and move up the graph to the temperature you recorded at 1 minute. Place a dot on the graph at that point. Plot each temperature in the same way. When you have plotted all of your data, connect the dots with a smooth line.

Analysis

7. Examine your graph. Do you think the water heated faster than it cooled? Explain.

8. Estimate what the temperature of the water was 2.5 minutes after you placed the beaker on the hot plate. Explain how you can make a good estimate of temperature between those you recorded.

9. Explain how a graph may give more information than the same data in a chart.

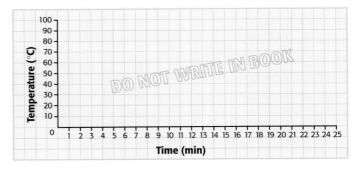

DO NOT WRITE IN BOOK

Datasheets for LabBook
Datasheet

Edith C. McAlanis
Socorro Middle School
El Paso, Texas

Answers

7. Answers will vary according to several factors, including altitude. Students should notice whether there is a gentle slope of the line (indicating gradual heating or cooling) or a steep slope (indicating rapid heating or cooling).

8. Answers will vary.

9. A list or a chart is organized information, and sometimes it is necessary to put collected data into one of these forms before graphing. Because a graph is like a picture, it can often help scientists to see what is happening when numbers alone would be confusing. A graph can show a trend or a pattern that may not be readily discernible in a list or chart.

A Window to a Hidden World

Have you ever noticed that objects underwater appear closer than they really are? That's because light waves change speed when they travel from air into water. Anton van Leeuwenhoek, a pioneer of microscopy in the late seventeenth century, used a drop of water to magnify objects. That drop of water brought a hidden world closer into view. How did Leeuwenhoek's microscope work? In this investigation, you will build a model of it to find out.

Procedure

1. Punch a hole in the center of the poster board with a hole punch, as shown in (a) at right.

2. Tape a small piece of clear plastic wrap over the hole, as shown in (b) at right. Be sure the plastic wrap is large enough so that the tape you use to secure it does not cover the hole.

3. Use an eyedropper to put one drop of water over the hole. Check to be sure your drop of water is dome-shaped (convex), as shown in (c) at right.

4. Hold the microscope close to your eye and look through the drop. Be careful not to disturb the water drop.

5. Hold the microscope over a piece of newspaper and observe the image.

Analysis

6. Describe and draw the image you see. Is the image larger or the same size as it was without the microscope? Is the image clear or blurred? Is the shape of the image distorted?

7. How do you think your model could be improved?

Going Further
Robert Hooke and Zacharias Janssen contributed much to the field of microscopy. Find out who they were, when they lived, and what they did.

Materials

- hole punch
- 3 × 10 cm piece of poster board
- clear plastic wrap
- transparent tape
- eyedropper
- water
- newspaper

(a)

(b)

(c)

Teacher's Notes sidebar

A Window to a Hidden World
Teacher's Notes

Time Required
One 45-minute class period

Lab Ratings

EASY ——————→ HARD

TEACHER PREP
STUDENT SET-UP
CONCEPT LEVEL
CLEAN UP

MATERIALS
The materials listed on the student page are enough for a group of 4–5 students. A 3 × 5 in. index card cut in half lengthwise can substitute for a stiff piece of poster board. It can be difficult to eliminate wrinkles in the plastic over the hole. Some students may need assistance.

Datasheets for LabBook
Datasheet

Science Skills Worksheet
"Using Models to Communicate"

Answers

6. Answers will vary. Students should see a slightly larger image. It will be blurred, especially around the edges. The image may be distorted.

7. Most students will think their model could be improved by eliminating the wrinkles over the hole.

Going Further

Robert Hooke (1635–1703), one of the world's great inventors, is famous for his discovery of "cells" in cork tissue as seen through his improved microscope. Hooke was also a keen observer with an interest in fossils and geology. Zacharias Janssen, a Dutch lens grinder, mounted two lenses in a tube to produce the first compound microscope in 1590.

Georgiann Delgadillo
East Valley School District
Continuous Curriculum School
Spokane, Washington

The Best-Bread Bakery Dilemma
Teacher's Notes

Time Required

Two 45-minute class periods

Lab Ratings

EASY —————→ HARD

TEACHER PREP 🧪🧪
STUDENT SET-UP 🧪
CONCEPT LEVEL 🧪
CLEAN UP 🧪

MATERIALS

The materials listed on the student page are enough for a group of 3–4 students. Yeast is easily obtained from the local grocery store. The school cafeteria may be willing to donate the amount you need.

You may wish to add other materials in anticipation of students' experimental design. For example, some students may recognize that they could collect CO_2 in a balloon attached to the top of a test tube containing live yeast.

Safety Caution

Remind students to review all safety cautions and icons before beginning this lab activity.

Caution students to be careful of the hot plate and the cord. You should demonstrate the proper laboratory technique for determining the presence of an odor. Hold the container away from your face about 25 cm and just below your nose. Use the other hand to "waft" the odor toward your face. Caution students NEVER to put their noses directly in a container and inhale.

The Best-Bread Bakery Dilemma

The chief baker at the Best-Bread Bakery thinks that the yeast the bakery received may be dead. Yeast is a central ingredient in bread. Yeast is a living organism, a member of the kingdom Fungi, and it undergoes the same life processes as other living organisms. When yeast grows in the presence of oxygen and other nutrients, it produces carbon dioxide. The gas forms bubbles that cause the dough to rise. Thousands of dollars may be lost if the yeast is dead.

The Best-Bread Bakery has requested that you test the yeast. The bakery has furnished samples of live yeast and some samples of the yeast in question.

Materials

- yeast samples (live, A, and B)
- magnifying lens
- test tubes or clear plastic cups
- test-tube rack
- 250 mL beaker
- 125 mL of water
- hot plate
- Celsius thermometer with clip
- scoopula or small spoon
- sugar
- graduated cylinder
- 3 wooden stirring sticks
- flour
- heat-resistant gloves

Procedure

1. Make a data table similar to the one below. Leave plenty of room to write your observations.

2. Examine each yeast sample with a magnifying lens. You may want to sniff the samples to determine the presence of an odor. (Your teacher will demonstrate the appropriate way to detect odors in the laboratory.) Record your observations in the data table.

3. Label three test tubes or plastic cups "Live Yeast," "Sample A Yeast," and "Sample B Yeast."

4. Fill a beaker with 125 mL of water, and place the beaker on a hot plate. Use a thermometer to be sure the water does not get warmer than 32°C. Attach the thermometer to the side of the beaker with a clip so the thermometer doesn't touch the bottom of the beaker. Turn off the hot plate when the temperature reaches 32°C.

	Observations	0 min	5 min	10 min	15 min	20 min	25 min	Dead or alive?
Live yeast								
Sample A yeast								
Sample B yeast								

DO NOT WRITE IN BOOK

134

Datasheets for LabBook
Datasheet

Science Skills
Worksheet
"Science Writing"

Susan Gorman
North Ridge Middle School
North Richland Hills, Texas

5. Add a small scoop (about 1/2 tsp) of each yeast sample to the correctly labeled container. Add a small scoop of sugar to each container.

6. Add 10 mL of the warm water to each container, and stir.

7. Add a small scoop of flour to each container, and stir again. The flour will help make the process more visible but is not necessary as food for the yeast.

8. Observe the samples carefully. Look for bubbles. Make observations at 5-minute intervals. Write your observations in the data table.

9. In the last column of the data table, write "alive" or "dead" based on your observations during the experiment.

Analysis

10. Describe any differences in the yeast samples before the experiment.

11. Describe the appearance of the yeast samples at the conclusion of the experiment.

12. Why was a sample of live yeast included in the experiment?

13. Why was sugar added to the samples?

14. Based on your observations, is either sample alive?

15. Write a letter to the Best-Bread Bakery stating your recommendation to use or not use the yeast samples. Give reasons for your recommendation.

Going Further

Based on your observations of the nutrient requirements of yeast, design an experiment to determine the ideal combination of nutrients. Vary the amount of nutrients or examine different energy sources.

Preparation Notes

At least one of the suspect samples should be killed yeast. To do this, place the yeast in an oven at 400°F for 10 minutes or in a microwave oven for a few minutes on high. Do not allow yeast to become moist before use. Toothpicks, coffee stirrers, etc., may be used for stirring. The amounts of each ingredient used are not definite, and you may wish to vary amounts, depending on the results desired.

Lab Notes

To help students prepare for this activity, you may wish to review cellular respiration and fermentation. The equation for respiration is:

$C_6H_{12}O_6 + 6O_2 \rightarrow 6CO_2 + 6H_2O + energy$

135

Answers

10. Answers are based on students' observations and will vary.

11. Answers will vary.

12. Live yeast is included so students can observe bubble formation from the respiration of living organisms.

13. Sugar is added as a nutrient for the living yeast.

14. Answers will vary according to students' experimental protocol.

15. Student letters will vary, but should recommend the optimal samples they determined in their experiment.

Aunt Flossie and the Intruder
Teacher's Notes

Time Required

Three 20-minute brainstorming and design sessions and five 5-minute observation periods on successive days

Lab Ratings

EASY ——→ HARD

TEACHER PREP 🧪🧪
STUDENT SET-UP 🧪🧪🧪
CONCEPT LEVEL 🧪🧪🧪🧪
CLEAN UP 🧪🧪

MATERIALS

Students will need to submit a list of supplies and equipment they will need for their experiment to you for approval.

Safety Caution

Be sure students address any safety concerns in the design of their experiments.

 Datasheets for LabBook Datasheet

Elizabeth Rustad
Crane Junior High School
Yuma, Arizona

Aunt Flossie and the Intruder

Aunt Flossie is a *really* bad housekeeper! She *never* cleans the refrigerator, and things get really gross in there. Last week she pulled out a plastic resealable bag that looked like it was going to explode! The bag was full of gas that she did not put there! Aunt Flossie remembered from her school days that gases are released from living things as waste products. Something had to be alive in the bag! Aunt Flossie became very upset that there was an intruder in her refrigerator. She said she would not bake another cookie until you determine the nature of the intruder.

Materials

- items to be determined by the students and approved by teacher as needed for each experiment such as resealable plastic bags, food samples, a scale, or a thermometer.
- protective gloves

Procedure

1. Design an investigation to determine how gas got into Aunt Flossie's bag. In your ScienceLog, make a list of the materials you will need, and prepare all the data tables you will need for recording your observations.

2. As you design your investigation, be sure to include each of the steps listed at right.

3. Get your teacher's approval of your experimental design and your list of materials before you begin.

4. Dispose of your materials according to your teacher's instructions at the end of your experiment.
   Caution: Do not open any bags of spoiled food or allow any of the contents to escape.

- Ask a question
- Form a hypothesis
- Test the hypothesis
- Analyze the data
- Draw conclusions
- Communicate your results

Analysis

5. Write a letter to Aunt Flossie describing your experiment. Explain what produced the gas in the bag and your recommendations for preventing these intruders in her refrigerator in the future. Invite her to bake cookies for your class!

Going Further

Research in the library or on the Internet to find out how people kept food fresh before refrigeration. Find out what advances have been made in food preservation as a result of the space program.

136

Answers

5. Students should explain (in a letter to Aunt Flossie) that it was mold or bacteria in the baggie that produced the gas. The gas was a product of the respiration of the mold or bacteria on the food. Preventing future intruders could be as simple as lowering the temperature of the refrigerator or freezing food.

Going Further

Students should learn that people have developed many ways of preserving food. Salt and sugar have been used as "curing" agents for meat, and home canning of foods has been around for generations. The space program has given us many new technologies of food preservation, particularly freeze-drying.

Travelin' Seeds

SKILL BUILDER

You have learned from your study of plants that there are some very interesting and unusual plant adaptations. Some of the most interesting adaptations are modifications that allow plant seeds and fruits to be dispersed, or scattered, away from the parent plant. This dispersal enables the young seedlings to obtain the space, sun, and other resources they need without directly competing with the parent plant.

In this activity, you will use your own creativity to disperse a seed.

Procedure

1. Obtain a seed and a dispersal challenge card from your teacher. In your ScienceLog, record the type of challenge card you have been given.

2. Create a plan for using the available materials to disperse your seed as described on the challenge card. Record your plan in your ScienceLog. Get your teacher's approval before proceeding.

3. With your teacher's permission, test your seed-dispersal method. Perform several trials. Make a data table in your ScienceLog, and record the results of your trials.

Analysis

4. Were you able to successfully complete the seed-dispersal challenge? Explain.

5. Are there any modifications you could make to your method to improve the dispersal of your seed?

6. Describe some plants that disperse their seeds in a way similar to your seed-dispersal method.

Materials

- bean seed
- seed-dispersal challenge card
- various household or recycled materials (examples: glue, tape, paper, paper clips, rubber bands, cloth, paper cups and plates, paper towels, cardboard)

◀ Mangrove seed

◀ Cottonwood

Wild berry ▶

Grass bur ▶

Jane Lemons
Western Rockingham
Middle School
Madison, North Carolina

Datasheets for LabBook
Datasheet

Answers
4.–6. Answers will vary.

LabBook

Travelin' Seeds
Teacher's Notes

Time Required
One 45-minute class period

Lab Ratings

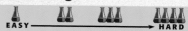
EASY ————————▶ HARD

TEACHER PREP 🧪🧪
STUDENT SET-UP 🧪🧪
CONCEPT LEVEL 🧪🧪
CLEAN UP 🧪

Preparation Notes

Challenge Cards Make up seed-dispersal challenge cards ahead of time. Five basic challenges are listed below. Copy each as needed so that there is one card per student or group.

1. Modify your seed so that it will be carried on an animal's fur as the animal passes by the plant. The seed must travel for at least 1 m. The animal must not be aware that the seed has been attached.

2. Modify your seed so that it will be flung 1 m away from the parent plant (you/group). Be careful that your tests do not fly at others in the class.

3. Modify your seed so that it will glide or float in the air when it falls off the parent plant. The seed must be dropped, not pushed or thrown, and must travel at least 1 m.

4. Modify your seed so that animals will find it desirable to eat and digest.

5. Modify your seed so that it will float on water for at least 1 minute.

Weepy Weeds
Teacher's Notes

Time Required

One or two 45-minute class periods

Lab Ratings

EASY —————→ HARD

TEACHER PREP ▲▲
STUDENT SET-UP ▲▲
CONCEPT LEVEL ▲▲
CLEAN UP ▲▲

MATERIALS

The materials listed on the student page are enough for 1 student. The plant used in this lab can be any leafy plant, such as a bean plant or a coleus. The plant shown is a coleus with all but the top four leaves trimmed away.

Safety Caution

Remind students to review all safety cautions and icons before beginning this lab activity.

Lab Notes

If your lab period is short, you may want to eliminate the measurement of the height of water in the test tube at 40 minutes.

Although it is not essential to the activity, you may want to begin with an exact amount of water in each test tube. Students would then know that the difference can be due only to evaporation and transpiration.

Weepy Weeds

You are trying to find a way to drain an area that is flooded with water polluted with fertilizer. You know that a plant releases water through the stomata in its leaves. As water evaporates from the leaves, more water is pulled up from the roots through the stem and into the leaves. By this process, called transpiration, water and nutrients are pulled into the plant from the soil. About 90 percent of the water a plant takes up through its roots is released into the atmosphere as water vapor through transpiration. Your idea is to add plants to the flooded area that will transpire the water and take up the fertilizer in their roots.

How much water can a plant take up and release in a certain period of time? In this activity, you will observe transpiration and determine one stem's rate of transpiration.

Materials

- 2 test tubes
- test-tube rack
- water
- coleus or other plant stem cutting
- glass-marking pen
- metric ruler
- clock
- graph paper

Procedure

1. In your ScienceLog, make a data table similar to the one below for recording your measurements.

Height of Water in Test Tubes		
Time	Test tube with plant	Test tube without plant
Initial		
After 10 min		
After 20 min		
After 30 min		
After 40 min		
Overnight		

DO NOT WRITE IN BOOK

2. Fill each test tube approximately three-fourths full of water. Place both test tubes in a test-tube rack.

3. Place the plant stem so that it stands upright in one of the test tubes. Your test tubes should look like the ones in the photograph at right.

4. Use the glass-marking pen to mark the water level in each of the test tubes. Be sure you have the plant stem in place in its test tube before you mark the water level. Why is this necessary?

 Datasheets for LabBook
Datasheet

 Science Skills Worksheet
"Interpreting Your Data"

CLASSROOM TESTED & APPROVED

David Sparks
Redwater Junior High School
Redwater, Texas

5. Measure the height of the water in each test tube. Be sure to hold the test tube level, and measure from the waterline to the bottom of the curve at the bottom of the test tube. Record these measurements on the row labeled "Initial."

6. Wait 10 minutes, and measure the height of the water in each test tube again. Record these measurements in your data table.

7. Repeat step 6 three more times. Record your measurements each time.

8. Wait 24 hours, and measure the height of the water in each test tube. Record these measurements in your data table.

9. Construct a graph similar to the one below. Plot the data from your data table. Draw a line for each test tube. Use a different color for each line, and make a key below your graph.

10. Calculate the rate of transpiration for your plant by using the following operations:

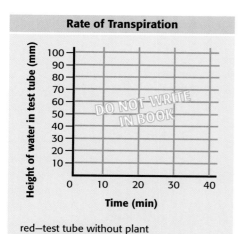

Rate of Transpiration

(graph: Height of water in test tube (mm) vs Time (min))

DO NOT WRITE IN BOOK

red—test tube without plant
blue—test tube with plant

Test tube with plant:
Initial height
− Overnight height
‾‾‾‾‾‾‾‾‾‾‾‾‾‾‾‾
Difference in height of water **(A)**

Test tube without plant:
Initial height
− Overnight height
‾‾‾‾‾‾‾‾‾‾‾‾‾‾‾‾
Difference in height of water **(B)**

Water height difference due to transpiration:
Difference **A**
− Difference **B**
‾‾‾‾‾‾‾‾‾‾‾‾‾‾‾‾
Water lost due to transpiration (in millimeters) in 24 hours

Analysis

11. What was the purpose of the test tube that held only water?

12. What caused the water to go down in the test tube containing the plant stem? Did the same thing happen in the test tube with water only? Explain your answer.

13. What was the calculated rate of transpiration per day?

14. Using your graph, compare the rate of transpiration with the rate of evaporation alone.

15. Prepare a presentation of your experiment for your class. Use your data tables, graphs, and calculations as visual aids.

Going Further

How many leaves did your plant sprigs have? Use this number to estimate what the rate of transpiration might be for a plant with 200 leaves. When you have your answer in millimeters of height in a test tube, pour this amount into a graduated cylinder to measure it in milliliters.

Answers

11. The test tube that holds only water is a control; it will lose water only by evaporation.

12. Water in the test tube containing the plant stem will be lost through evaporation and transpiration. Evaporation is the only means of water loss in the test tube without the plant stem.

13. This answer will vary according to several variables in the classroom, such as the amount of light and the temperature.

14. This answer will vary. Have students compare and contrast the lines on the graph and explain how the graph is easier to interpret than numbers in a data list.

Going Further

Answers will vary.

Concept Mapping: A Way to Bring Ideas Together

What Is a Concept Map?

Have you ever tried to tell someone about a book or a chapter you've just read and found that you can remember only a few isolated words and ideas? Or maybe you've memorized facts for a test and then weeks later discovered you're not even sure what topics those facts covered.

In both cases, you may have understood the ideas or concepts by themselves but not in relation to one another. If you could somehow link the ideas together, you would probably understand them better and remember them longer. This is something a concept map can help you do. A concept map is a way to see how ideas or concepts fit together. It can help you see the "big picture."

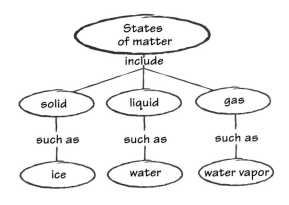

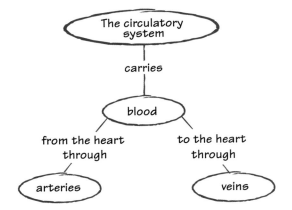

How to Make a Concept Map

1 **Make a list of the main ideas or concepts.**

It might help to write each concept on its own slip of paper. This will make it easier to rearrange the concepts as many times as necessary to make sense of how the concepts are connected. After you've made a few concept maps this way, you can go directly from writing your list to actually making the map.

2 **Arrange the concepts in order from the most general to the most specific.**

Put the most general concept at the top and circle it. Ask yourself, "How does this concept relate to the remaining concepts?" As you see the relationships, arrange the concepts in order from general to specific.

3 **Connect the related concepts with lines.**

4 **On each line, write an action word or short phrase that shows how the concepts are related.**

Look at the concept maps on this page, and then see if you can make one for the following terms:

plants, water, photosynthesis, carbon dioxide, sun's energy

One possible answer is provided at right, but don't look at it until you try the concept map yourself.

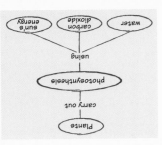

SI Measurement

The International System of Units, or SI, is the standard system of measurement used by many scientists. Using the same standards of measurement makes it easier for scientists to communicate with one another.

SI works by combining prefixes and base units. Each base unit can be used with different prefixes to define smaller and larger quantities. The table below lists common SI prefixes.

SI Prefixes			
Prefix	**Abbreviation**	**Factor**	**Example**
kilo-	k	1,000	kilogram, 1 kg = 1,000 g
hecto-	h	100	hectoliter, 1 hL = 100 L
deka-	da	10	dekameter, 1 dam = 10 m
		1	meter, liter
deci-	d	0.1	decigram, 1 dg = 0.1 g
centi-	c	0.01	centimeter, 1 cm = 0.01 m
milli-	m	0.001	milliliter, 1 mL = 0.001 L
micro-	µ	0.000 001	micrometer, 1 µm = 0.000 001 m

SI Conversion Table		
SI units	**From SI to English**	**From English to SI**
Length		
kilometer (km) = 1,000 m	1 km = 0.621 mi	1 mi = 1.609 km
meter (m) = 100 cm	1 m = 3.281 ft	1 ft = 0.305 m
centimeter (cm) = 0.01 m	1 cm = 0.394 in.	1 in. = 2.540 cm
millimeter (mm) = 0.001 m	1 mm = 0.039 in.	
micrometer (µm) = 0.000 001 m		
nanometer (nm) = 0.000 000 001 m		
Area		
square kilometer (km^2) = 100 hectares	1 km^2 = 0.386 mi^2	1 mi^2 = 2.590 km^2
hectare (ha) = 10,000 m^2	1 ha = 2.471 acres	1 acre = 0.405 ha
square meter (m^2) = 10,000 cm^2	1 m^2 = 10.765 ft^2	1 ft^2 = 0.093 m^2
square centimeter (cm^2) = 100 mm^2	1 cm^2 = 0.155 in.^{2}	1 in.^{2} = 6.452 cm^2
Volume		
liter (L) = 1,000 mL = 1 dm^3	1 L = 1.057 fl qt	1 fl qt = 0.946 L
milliliter (mL) = 0.001 L = 1 cm^3	1 mL = 0.034 fl oz	1 fl oz = 29.575 mL
microliter (µL) = 0.000 001 L		
Mass		
kilogram (kg) = 1,000 g	1 kg = 2.205 lb	1 lb = 0.454 kg
gram (g) = 1,000 mg	1 g = 0.035 oz	1 oz = 28.349 g
milligram (mg) = 0.001 g		
microgram (µg) = 0.000 001 g		

Temperature Scales

Temperature can be expressed using three different scales: Fahrenheit, Celsius, and Kelvin. The SI unit for temperature is the kelvin (K).

Although 0 K is much colder than 0°C, a change of 1 K is equal to a change of 1°C.

Three Temperature Scales

	Fahrenheit	Celsius	Kelvin
Water boils	212°	100°	373
Body temperature	98.6°	37°	310
Room temperature	68°	20°	293
Water freezes	32°	0°	273

Temperature Conversions Table

To convert	Use this equation:	Example
Celsius to Fahrenheit °C ⟶ °F	$°F = \left(\frac{9}{5} \times °C\right) + 32$	Convert 45°C to °F. $°F = \left(\frac{9}{5} \times 45°C\right) + 32 = 113°F$
Fahrenheit to Celsius °F ⟶ °C	$°C = \frac{5}{9} \times (°F - 32)$	Convert 68°F to °C. $°C = \frac{5}{9} \times (68°F - 32) = 20°C$
Celsius to Kelvin °C ⟶ K	$K = °C + 273$	Convert 45°C to K. $K = 45°C + 273 = 318\ K$
Kelvin to Celsius K ⟶ °C	$°C = K - 273$	Convert 32 K to °C. $°C = 32\ K - 273 = -241°C$

Measuring Skills

Using a Graduated Cylinder

When using a graduated cylinder to measure volume, keep the following procedures in mind:

1 Make sure the cylinder is on a flat, level surface.

2 Move your head so that your eye is level with the surface of the liquid.

3 Read the mark closest to the liquid level. On glass graduated cylinders, read the mark closest to the center of the curve in the liquid's surface.

Using a Meterstick or Metric Ruler

When using a meterstick or metric ruler to measure length, keep the following procedures in mind:

1 Place the ruler firmly against the object you are measuring.

2 Align one edge of the object exactly with the zero end of the ruler.

3 Look at the other edge of the object to see which of the marks on the ruler is closest to that edge. **Note:** Each small slash between the centimeters represents a millimeter, which is one-tenth of a centimeter.

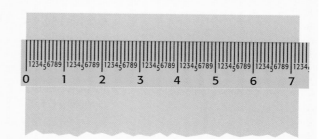

Using a Triple-Beam Balance

When using a triple-beam balance to measure mass, keep the following procedures in mind:

1 Make sure the balance is on a level surface.

2 Place all of the countermasses at zero. Adjust the balancing knob until the pointer rests at zero.

3 Place the object you wish to measure on the pan. **Caution:** Do not place hot objects or chemicals directly on the balance pan.

4 Move the largest countermass along the beam to the right until it is at the last notch that does not tip the balance. Follow the same procedure with the next-largest countermass. Then move the smallest countermass until the pointer rests at zero.

5 Add the readings from the three beams together to determine the mass of the object.

6 When determining the mass of crystals or powders, use a piece of filter paper. First find the mass of the paper. Then add the crystals or powder to the paper and re-measure. The actual mass of the crystals or powder is the total mass minus the mass of the paper. When finding the mass of liquids, first find the mass of the empty container. Then find the mass of the liquid and container together. The mass of the liquid is the total mass minus the mass of the container.

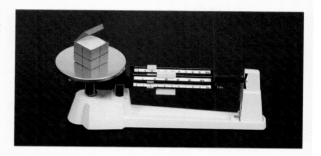

Scientific Method

The series of steps that scientists use to answer questions and solve problems is often called the **scientific method.** The scientific method is not a rigid procedure. Scientists may use all of the steps or just some of the steps of the scientific method. They may even repeat some of the steps. The goal of the scientific method is to come up with reliable answers and solutions.

Six Steps of the Scientific Method

1 **Ask a Question** Good questions come from careful **observations.** You make observations by using your senses to gather information. Sometimes you may use instruments, such as microscopes and telescopes, to extend the range of your senses. As you observe the natural world, you will discover that you have many more questions than answers. These questions drive the scientific method.

Questions beginning with *what, why, how,* and *when* are very important in focusing an investigation, and they often lead to a hypothesis. (You will learn what a hypothesis is in the next step.) Here is an example of a question that could lead to further investigation.

Question: How does acid rain affect plant growth?

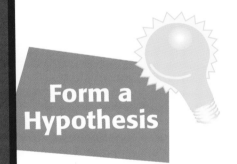

2 **Form a Hypothesis** After you come up with a question, you need to turn the question into a **hypothesis.** A hypothesis is a clear statement of what you expect the answer to your question to be. Your hypothesis will represent your best "educated guess" based on your observations and what you already know. A good hypothesis is testable. If observations and information cannot be gathered or if an experiment cannot be designed to test your hypothesis, it is untestable, and the investigation can go no further.

Here is a hypothesis that could be formed from the question, "How does acid rain affect plant growth?"

Hypothesis: Acid rain causes plants to grow more slowly.

Notice that the hypothesis provides some specifics that lead to methods of testing. The hypothesis can also lead to predictions. A **prediction** is what you think will be the outcome of your experiment or data collection. Predictions are usually stated in an "if . . . then" format. For example, **if** meat is kept at room temperature, **then** it will spoil faster than meat kept in the refrigerator. More than one prediction can be made for a single hypothesis. Here is a sample prediction for the hypothesis that acid rain causes plants to grow more slowly.

Prediction: If a plant is watered with only acid rain (which has a pH of 4), then the plant will grow at half its normal rate.

3 **Test the Hypothesis** After you have formed a hypothesis and made a prediction, you should test your hypothesis. There are different ways to do this. Perhaps the most familiar way is to conduct a **controlled experiment.** A controlled experiment tests only one factor at a time. A controlled experiment has a **control group** and one or more **experimental groups.** All the factors for the control and experimental groups are the same except for one factor, which is called the **variable.** By changing only one factor, you can see the results of just that one change.

Sometimes, the nature of an investigation makes a controlled experiment impossible. For example, dinosaurs have been extinct for millions of years, and the Earth's core is surrounded by thousands of meters of rock. It would be difficult, if not impossible, to conduct controlled experiments on such things. Under such circumstances, a hypothesis may be tested by making detailed observations. Taking measurements is one way of making observations.

Test the Hypothesis

4 **Analyze the Results** After you have completed your experiments, made your observations, and collected your data, you must analyze all the information you have gathered. Tables and graphs are often used in this step to organize the data.

Analyze the Results

5 **Draw Conclusions** Based on the analysis of your data, you should conclude whether or not your results support your hypothesis. If your hypothesis is supported, you (or others) might want to repeat the observations or experiments to verify your results. If your hypothesis is not supported by the data, you may have to check your procedure for errors. You may even have to reject your hypothesis and make a new one. If you cannot draw a conclusion from your results, you may have to try the investigation again or carry out further observations or experiments.

Draw Conclusions

No

Do they support your hypothesis?

Yes

6 **Communicate Results** After any scientific investigation, you should report your results. By doing a written or oral report, you let others know what you have learned. They may want to repeat your investigation to see if they get the same results. Your report may even lead to another question, which in turn may lead to another investigation.

Communicate Results

Scientific Method in Action

The scientific method is not a "straight line" of steps. It contains loops in which several steps may be repeated over and over again, while others may not be necessary. For example, sometimes scientists will find that testing one hypothesis raises new questions and new hypotheses to be tested. And sometimes, testing the hypothesis leads directly to a conclusion. Furthermore, the steps in the scientific method are not always used in the same order. Follow the steps in the diagram below, and see how many different directions the scientific method can take you.

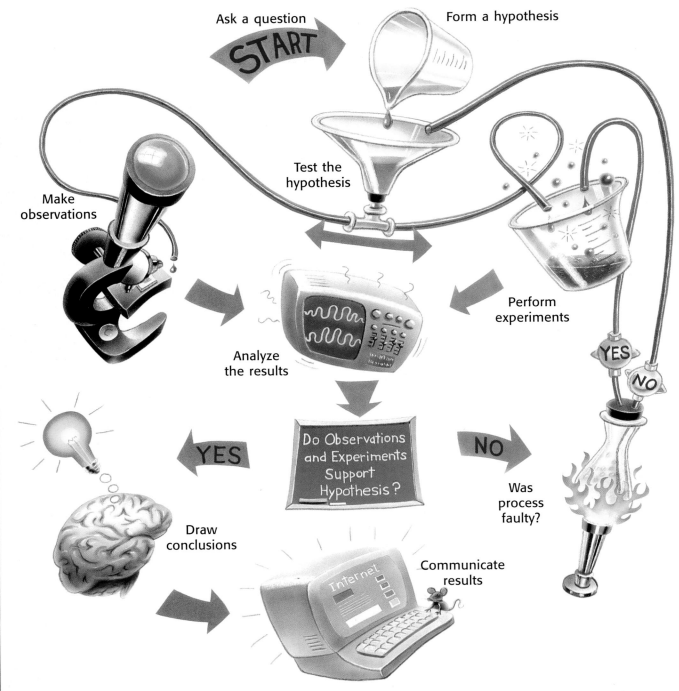

Ask a question

START

Form a hypothesis

Test the hypothesis

Make observations

Perform experiments

Analyze the results

YES

NO

YES

Do Observations and Experiments Support Hypothesis?

NO

Was process faulty?

Draw conclusions

Communicate results

Internet

Making Charts and Graphs

Circle Graphs

A circle graph, or pie chart, shows how each group of data relates to all of the data. Each part of the circle represents a category of the data. The entire circle represents all of the data. For example, a biologist studying a hardwood forest in Wisconsin found that there were five different types of trees. The data table at right summarizes the biologist's findings.

Wisconsin Hardwood Trees	
Type of tree	**Number found**
Oak	600
Maple	750
Beech	300
Birch	1,200
Hickory	150
Total	3,000

How to Make a Circle Graph

1 In order to make a circle graph of this data, first find the percentage of each type of tree. To do this, divide the number of individual trees by the total number of trees and multiply by 100.

$$\frac{600 \text{ oak}}{3,000 \text{ trees}} \times 100 = 20\%$$

$$\frac{750 \text{ maple}}{3,000 \text{ trees}} \times 100 = 25\%$$

$$\frac{300 \text{ beech}}{3,000 \text{ trees}} \times 100 = 10\%$$

$$\frac{1,200 \text{ birch}}{3,000 \text{ trees}} \times 100 = 40\%$$

$$\frac{150 \text{ hickory}}{3,000 \text{ trees}} \times 100 = 5\%$$

2 Now determine the size of the pie shapes that make up the chart. Do this by multiplying each percentage by 360°. Remember that a circle contains 360°.

$20\% \times 360° = 72°$ $25\% \times 360° = 90°$
$10\% \times 360° = 36°$ $40\% \times 360° = 144°$
$5\% \times 360° = 18°$

3 Then check that the sum of the percentages is 100 and the sum of the degrees is 360.

$20\% + 25\% + 10\% + 40\% + 5\% = 100\%$
$72° + 90° + 36° + 144° + 18° = 360°$

4 Use a compass to draw a circle and mark its center.

5 Then use a protractor to draw angles of 72°, 90°, 36°, 144°, and 18° in the circle.

6 Finally, label each part of the graph, and choose an appropriate title.

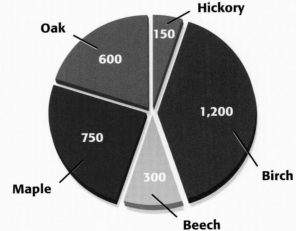

A Community of Wisconsin Hardwood Trees

Hickory 150 · Oak 600 · Birch 1,200 · Maple 750 · Beech 300

Line Graphs

Population of Appleton, 1900–2000	
Year	Population
1900	1,800
1920	2,500
1940	3,200
1960	3,900
1980	4,600
2000	5,300

Line graphs are most often used to demonstrate continuous change. For example, Mr. Smith's science class analyzed the population records for their hometown, Appleton, between 1900 and 2000. Examine the data at left.

Because the year and the population change, they are the *variables*. The population is determined by, or dependent on, the year. Therefore, the population is called the **dependent variable**, and the year is called the **independent variable**. Each set of data is called a **data pair**. To prepare a line graph, data pairs must first be organized in a table like the one at left.

How to Make a Line Graph

1 Place the independent variable along the horizontal (*x*) axis. Place the dependent variable along the vertical (*y*) axis.

2 Label the *x*-axis "Year" and the *y*-axis "Population." Look at your largest and smallest values for the population. Determine a scale for the *y*-axis that will provide enough space to show these values. You must use the same scale for the entire length of the axis. Find an appropriate scale for the *x*-axis too.

3 Choose reasonable starting points for each axis.

4 Plot the data pairs as accurately as possible.

5 Choose a title that accurately represents the data.

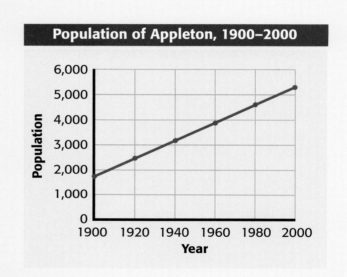

How to Determine Slope

Slope is the ratio of the change in the *y*-axis to the change in the *x*-axis, or "rise over run."

1 Choose two points on the line graph. For example, the population of Appleton in 2000 was 5,300 people. Therefore, you can define point *a* as (2000, 5,300). In 1900, the population was 1,800 people. Define point *b* as (1900, 1,800).

2 Find the change in the *y*-axis.
(*y* at point *a*) − (*y* at point *b*)
5,300 people − 1,800 people = 3,500 people

3 Find the change in the *x*-axis.
(*x* at point *a*) − (*x* at point *b*)
2000 − 1900 = 100 years

4 Calculate the slope of the graph by dividing the change in *y* by the change in *x*.

$$\text{slope} = \frac{\text{change in } y}{\text{change in } x}$$

$$\text{slope} = \frac{3{,}500 \text{ people}}{100 \text{ years}}$$

slope = 35 people per year

In this example, the population in Appleton increased by a fixed amount each year. The graph of this data is a straight line. Therefore, the relationship is **linear.** When the graph of a set of data is not a straight line, the relationship is **nonlinear.**

Using Algebra to Determine Slope

The equation in step 4 may also be arranged to be:

$$y = kx$$

where y represents the change in the y-axis, k represents the slope, and x represents the change in the x-axis.

$$\text{slope} = \frac{\text{change in } y}{\text{change in } x}$$

$$k = \frac{y}{x}$$

$$k \times x = \frac{y \times x}{x}$$

$$kx = y$$

Bar Graphs

Bar graphs are used to demonstrate change that is not continuous. These graphs can be used to indicate trends when the data are taken over a long period of time. A meteorologist gathered the precipitation records at right for Hartford, Connecticut, for April 1–15, 1996, and used a bar graph to represent the data.

Precipitation in Hartford, Connecticut April 1–15, 1996

Date	Precipitation (cm)	Date	Precipitation (cm)
April 1	0.5	April 9	0.25
April 2	1.25	April 10	0.0
April 3	0.0	April 11	1.0
April 4	0.0	April 12	0.0
April 5	0.0	April 13	0.25
April 6	0.0	April 14	0.0
April 7	0.0	April 15	6.50
April 8	1.75		

How to Make a Bar Graph

1 Use an appropriate scale and a reasonable starting point for each axis.

2 Label the axes, and plot the data.

3 Choose a title that accurately represents the data.

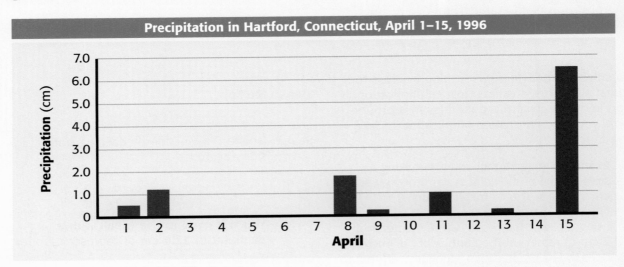

Appendix **149**

Math Refresher

Science requires an understanding of many math concepts. The following pages will help you review some important math skills.

Averages

An **average,** or **mean,** simplifies a list of numbers into a single number that *approximates* their value.

Example: Find the average of the following set of numbers: 5, 4, 7, and 8.

Step 1: Find the sum.

$$5 + 4 + 7 + 8 = 24$$

Step 2: Divide the sum by the amount of numbers in your set. Because there are four numbers in this example, divide the sum by 4.

$$\frac{24}{4} = 6$$

The average, or mean, is **6.**

Ratios

A **ratio** is a comparison between numbers, and it is usually written as a fraction.

Example: Find the ratio of thermometers to students if you have 36 thermometers and 48 students in your class.

Step 1: Make the ratio.

$$\frac{36 \text{ thermometers}}{48 \text{ students}}$$

Step 2: Reduce the fraction to its simplest form.

$$\frac{36}{48} = \frac{36 \div 12}{48 \div 12} = \frac{3}{4}$$

The ratio of thermometers to students is **3 to 4,** or $\frac{3}{4}$. The ratio may also be written in the form 3:4.

Proportions

A **proportion** is an equation that states that two ratios are equal.

$$\frac{3}{1} = \frac{12}{4}$$

To solve a proportion, first multiply across the equal sign. This is called cross-multiplication. If you know three of the quantities in a proportion, you can use cross-multiplication to find the fourth.

Example: Imagine that you are making a scale model of the solar system for your science project. The diameter of Jupiter is 11.2 times the diameter of the Earth. If you are using a plastic-foam ball with a diameter of 2 cm to represent the Earth, what diameter does the ball representing Jupiter need to be?

$$\frac{11.2}{1} = \frac{x}{2 \text{ cm}}$$

Step 1: Cross-multiply.

$$\frac{11.2}{1} \diagup\!\!\!\!\diagdown \frac{x}{2}$$

$$11.2 \times 2 = x \times 1$$

Step 2: Multiply.

$$22.4 = x \times 1$$

Step 3: Isolate the variable by dividing both sides by 1.

$$x = \frac{22.4}{1}$$

$$x = 22.4 \text{ cm}$$

You will need to use a ball with a diameter of **22.4 cm** to represent Jupiter.

Percentages

A **percentage** is a ratio of a given number to 100.

> **Example:** What is 85 percent of 40?

Step 1: Rewrite the percentage by moving the decimal point two places to the left.

$$.85$$

Step 2: Multiply the decimal by the number you are calculating the percentage of.

$$0.85 \times 40 = 34$$

85 percent of 40 is **34.**

Decimals

To **add** or **subtract decimals,** line up the digits vertically so that the decimal points line up. Then add or subtract the columns from right to left, carrying or borrowing numbers as necessary.

> **Example:** Add the following numbers: 3.1415 and 2.96.

Step 1: Line up the digits vertically so that the decimal points line up.

$$\begin{array}{r} 3.1415 \\ + \ 2.96 \\ \hline \end{array}$$

Step 2: Add the columns from right to left, carrying when necessary.

$$\begin{array}{r} 1 \ 1 \\ 3.1415 \\ + \ 2.96 \\ \hline 6.1015 \end{array}$$

The sum is **6.1015.**

Fractions

Numbers tell you how many; **fractions** tell you *how much of a whole.*

> **Example:** Your class has 24 plants. Your teacher instructs you to put 5 in a shady spot. What fraction does this represent?

Step 1: Write a fraction with the total number of parts in the whole as the denominator.

$$\frac{?}{24}$$

Step 2: Write the number of parts of the whole being represented as the numerator.

$$\frac{5}{24}$$

$\frac{5}{24}$ of the plants will be in the shade.

Reducing Fractions

It is usually best to express a fraction in simplest form. This is called *reducing* a fraction.

> **Example:** Reduce the fraction $\frac{30}{45}$ to its simplest form.

Step 1: Find the largest whole number that will divide evenly into both the numerator and denominator. This number is called the greatest common factor (GCF).

factors of the numerator 30: 1, 2, 3, 5, 6, 10, **15,** 30

factors of the denominator 45: 1, 3, 5, 9, **15,** 45

Step 2: Divide both the numerator and the denominator by the GCF, which in this case is 15.

$$\frac{30}{45} = \frac{30 \div 15}{45 \div 15} = \frac{2}{3}$$

$\frac{30}{45}$ reduced to its simplest form is $\frac{2}{3}$.

Adding and Subtracting Fractions

To **add** or **subtract fractions** that have the **same denominator,** simply add or subtract the numerators.

Examples:

$$\frac{3}{5} + \frac{1}{5} = ? \text{ and } \frac{3}{4} - \frac{1}{4} = ?$$

Step 1: Add or subtract the numerators.

$$\frac{3}{5} + \frac{1}{5} = \frac{4}{} \text{ and } \frac{3}{4} - \frac{1}{4} = \frac{2}{}$$

Step 2: Write the sum or difference over the denominator.

$$\frac{3}{5} + \frac{1}{5} = \frac{4}{5} \text{ and } \frac{3}{4} - \frac{1}{4} = \frac{2}{4}$$

Step 3: If necessary, reduce the fraction to its simplest form.

$$\frac{\mathbf{4}}{\mathbf{5}} \text{ cannot be reduced, and } \frac{2}{4} = \frac{\mathbf{1}}{\mathbf{2}}.$$

To **add** or **subtract fractions** that have **different denominators,** first find the least common denominator (LCD).

Examples:

$$\frac{1}{2} + \frac{1}{6} = ? \text{ and } \frac{3}{4} - \frac{2}{3} = ?$$

Step 1: Write the equivalent fractions with a common denominator.

$$\frac{3}{6} + \frac{1}{6} = ? \text{ and } \frac{9}{12} - \frac{8}{12} = ?$$

Step 2: Add or subtract.

$$\frac{3}{6} + \frac{1}{6} = \frac{4}{6} \text{ and } \frac{9}{12} - \frac{8}{12} = \frac{1}{12}$$

Step 3: If necessary, reduce the fraction to its simplest form.

$$\frac{4}{6} = \frac{\mathbf{2}}{\mathbf{3}}, \text{ and } \frac{\mathbf{1}}{\mathbf{12}} \text{ cannot be reduced.}$$

Multiplying Fractions

To **multiply fractions,** multiply the numerators and the denominators together, and then reduce the fraction to its simplest form.

Example:

$$\frac{5}{9} \times \frac{7}{10} = ?$$

Step 1: Multiply the numerators and denominators.

$$\frac{5}{9} \times \frac{7}{10} = \frac{5 \times 7}{9 \times 10} = \frac{35}{90}$$

Step 2: Reduce.

$$\frac{35}{90} = \frac{35 \div 5}{90 \div 5} = \frac{\mathbf{7}}{\mathbf{18}}$$

Dividing Fractions

To **divide fractions,** first rewrite the divisor (the number you divide *by*) upside down. This is called the reciprocal of the divisor. Then you can multiply and reduce if necessary.

Example:

$$\frac{5}{8} \div \frac{3}{2} = ?$$

Step 1: Rewrite the divisor as its reciprocal.

$$\frac{3}{2} \rightarrow \frac{2}{3}$$

Step 2: Multiply.

$$\frac{5}{8} \times \frac{2}{3} = \frac{5 \times 2}{8 \times 3} = \frac{10}{24}$$

Step 3: Reduce.

$$\frac{10}{24} = \frac{10 \div 2}{24 \div 2} = \frac{\mathbf{5}}{\mathbf{12}}$$

Scientific Notation

Scientific notation is a short way of representing very large and very small numbers without writing all of the place-holding zeros.

Example: Write 653,000,000 in scientific notation.

Step 1: Write the number without the place-holding zeros.

653

Step 2: Place the decimal point after the first digit.

6.53

Step 3: Find the exponent by counting the number of places that you moved the decimal point.

6.53000000

The decimal point was moved eight places to the left. Therefore, the exponent of 10 is positive 8. Remember, if the decimal point had moved to the right, the exponent would be negative.

Step 4: Write the number in scientific notation.

$$6.53 \times 10^8$$

Area

Area is the number of square units needed to cover the surface of an object.

Formulas:
Area of a square = side × side
Area of a rectangle = length × width
Area of a triangle = $\frac{1}{2}$ × base × height

Examples: Find the areas.

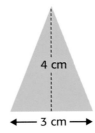

Triangle
Area = $\frac{1}{2}$ × base × height
Area = $\frac{1}{2}$ × 3 cm × 4 cm
Area = **6 cm²**

4 cm
3 cm

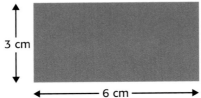

3 cm
6 cm

Rectangle
Area = length × width
Area = 6 cm × 3 cm
Area = **18 cm²**

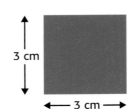

3 cm
3 cm

Square
Area = side × side
Area = 3 cm × 3 cm
Area = **9 cm²**

Volume

Volume is the amount of space something occupies.

Formulas:
Volume of a cube = side × side × side

Volume of a prism = area of base × height

Examples:
Find the volume of the solids.

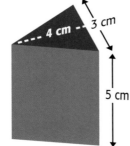

4 cm
3 cm
5 cm

Cube
Volume = side × side × side
Volume = 4 cm × 4 cm × 4 cm
Volume = **64 cm³**

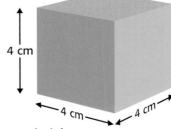

4 cm
4 cm
4 cm

Prism
Volume = area of base × height
Volume = (area of triangle) × height
Volume = $\left(\frac{1}{2} \times 3 \text{ cm} \times 4 \text{ cm}\right) \times 5 \text{ cm}$
Volume = 6 cm² × 5 cm
Volume = **30 cm³**

Periodic Table of the Elements

Each square on the table includes an element's name, chemical symbol, atomic number, and atomic mass.

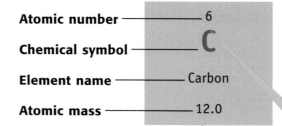

Atomic number ——— 6
Chemical symbol ——— C
Element name ——— Carbon
Atomic mass ——— 12.0

The background color indicates the type of element. Carbon is a nonmetal.

The color of the chemical symbol indicates the physical state at room temperature. Carbon is a solid.

Background
Metals
Metalloids
Nonmetals

Chemical Symbol
Solid
Liquid
Gas

Period 1

1
H
Hydrogen
1.0

	Group 1	Group 2	Group 3	Group 4	Group 5	Group 6	Group 7	Group 8	Group 9
Period 2	3 **Li** Lithium 6.9	4 **Be** Beryllium 9.0							
Period 3	11 **Na** Sodium 23.0	12 **Mg** Magnesium 24.3							
Period 4	19 **K** Potassium 39.1	20 **Ca** Calcium 40.1	21 **Sc** Scandium 45.0	22 **Ti** Titanium 47.9	23 **V** Vanadium 50.9	24 **Cr** Chromium 52.0	25 **Mn** Manganese 54.9	26 **Fe** Iron 55.8	27 **Co** Cobalt 58.9
Period 5	37 **Rb** Rubidium 85.5	38 **Sr** Strontium 87.6	39 **Y** Yttrium 88.9	40 **Zr** Zirconium 91.2	41 **Nb** Niobium 92.9	42 **Mo** Molybdenum 95.9	43 **Tc** Technetium (97.9)	44 **Ru** Ruthenium 101.1	45 **Rh** Rhodium 102.9
Period 6	55 **Cs** Cesium 132.9	56 **Ba** Barium 137.3	57 **La** Lanthanum 138.9	72 **Hf** Hafnium 178.5	73 **Ta** Tantalum 180.9	74 **W** Tungsten 183.8	75 **Re** Rhenium 186.2	76 **Os** Osmium 190.2	77 **Ir** Iridium 192.2
Period 7	87 **Fr** Francium (223.0)	88 **Ra** Radium (226.0)	89 **Ac** Actinium (227.0)	104 **Rf** Rutherfordium (261.1)	105 **Db** Dubnium (262.1)	106 **Sg** Seaborgium (263.1)	107 **Bh** Bohrium (262.1)	108 **Hs** Hassium (265)	109 **Mt** Meitnerium (266)

A row of elements is called a period.

A column of elements is called a group or family.

Lanthanides	58 **Ce** Cerium 140.1	59 **Pr** Praseodymium 140.9	60 **Nd** Neodymium 144.2	61 **Pm** Promethium (144.9)	62 **Sm** Samarium 150.4
Actinides	90 **Th** Thorium 232.0	91 **Pa** Protactinium 231.0	92 **U** Uranium 238.0	93 **Np** Neptunium (237.0)	94 **Pu** Plutonium 244.1

These elements are placed below the table to allow the table to be narrower.

This zigzag line reminds you where the metals, nonmetals, and metalloids are.

				Group 13	**Group 14**	**Group 15**	**Group 16**	**Group 17**	**Group 18**
									2 **He** Helium 4.0
				5 **B** Boron 10.8	6 **C** Carbon 12.0	7 **N** Nitrogen 14.0	8 **O** Oxygen 16.0	9 **F** Fluorine 19.0	10 **Ne** Neon 20.2
Group 10	**Group 11**	**Group 12**		13 **Al** Aluminum 27.0	14 **Si** Silicon 28.1	15 **P** Phosphorus 31.0	16 **S** Sulfur 32.1	17 **Cl** Chlorine 35.5	18 **Ar** Argon 39.9
28 **Ni** Nickel 58.7	29 **Cu** Copper 63.5	30 **Zn** Zinc 65.4	31 **Ga** Gallium 69.7	32 **Ge** Germanium 72.6	33 **As** Arsenic 74.9	34 **Se** Selenium 79.0	35 **Br** Bromine 79.9	36 **Kr** Krypton 83.8
46 **Pd** Palladium 106.4	47 **Ag** Silver 107.9	48 **Cd** Cadmium 112.4	49 **In** Indium 114.8	50 **Sn** Tin 118.7	51 **Sb** Antimony 121.8	52 **Te** Tellurium 127.6	53 **I** Iodine 126.9	54 **Xe** Xenon 131.3
78 **Pt** Platinum 195.1	79 **Au** Gold 197.0	80 **Hg** Mercury 200.6	81 **Tl** Thallium 204.4	82 **Pb** Lead 207.2	83 **Bi** Bismuth 209.0	84 **Po** Polonium (209.0)	85 **At** Astatine (210.0)	86 **Rn** Radon (222.0)
110 **Uun*** Ununnilium (271)	111 **Uuu*** Unununium (272)	112 **Uub*** Ununbium (277)		114 **Uuq*** Ununquadium (285)		116 **Uuh*** Ununhexium (289)		118 **Uuo*** Ununoctium (293)

A number in parenthesis is the mass number of the most stable form of that element.

63 **Eu** Europium 152.0	64 **Gd** Gadolinium 157.3	65 **Tb** Terbium 158.9	66 **Dy** Dysprosium 162.5	67 **Ho** Holmium 164.9	68 **Er** Erbium 167.3	69 **Tm** Thulium 168.9	70 **Yb** Ytterbium 173.0	71 **Lu** Lutetium 175.0
95 **Am** Americium (243.1)	96 **Cm** Curium (247.1)	97 **Bk** Berkelium (247.1)	98 **Cf** Californium (251.1)	99 **Es** Einsteinium (252.1)	100 **Fm** Fermium (257.1)	101 **Md** Mendelevium (258.1)	102 **No** Nobelium (259.1)	103 **Lr** Lawrencium (262.1)

*The official names and symbols for the elements greater than 109 will eventually be approved by a committee of scientists.

The Six Kingdoms

Kingdom Archaebacteria

The organisms in this kingdom are single-celled prokaryotes.

Archaebacteria		
Group	**Examples**	**Characteristics**
Methanogens	*Methanococcus*	found in soil, swamps, the digestive tract of mammals; produce methane gas; can't live in oxygen
Thermophiles	*Sulpholobus*	found in extremely hot environments; require sulphur, can't live in oxygen
Halophiles	*Halococcus*	found in environments with very high salt content, such as the Dead Sea; nearly all can live in oxygen

Kingdom Eubacteria

There are more than 4,000 named species in this kingdom of single-celled prokaryotes.

Eubacteria		
Group	**Examples**	**Characteristics**
Bacilli	*Escherichia coli*	rod-shaped; free-living, symbiotic, or parasitic; some can fix nitrogen; some cause disease
Cocci	*Streptococcus*	spherical-shaped, disease-causing; can form spores to resist unfavorable environments
Spirilla	*Treponema*	spiral-shaped; responsible for several serious illnesses, such as syphilis and Lyme disease

Kingdom Protista

The organisms in this kingdom are eukaryotes. There are single-celled and multicellular representatives.

Protists		
Group	**Examples**	**Characteristics**
Sacodines	*Amoeba*	radiolarians; single-celled consumers
Ciliates	*Paramecium*	single-celled consumers
Flagellates	*Trypanosoma*	single-celled parasites
Sporozoans	*Plasmodium*	single-celled parasites
Euglenas	*Euglena*	single-celled; photosynthesize
Diatoms	*Pinnularia*	most are single-celled; photosynthesize
Dinoflagellates	*Gymnodinium*	single-celled; some photosynthesize
Algae	*Volvox*, coral algae	4 phyla; single- or many-celled; photosynthesize
Slime molds	*Physarum*	single- or many-celled; consumers or decomposers
Water molds	powdery mildew	single- or many-celled, parasites or decomposers

Kingdom Fungi

There are single-celled and multicellular eukaryotes in this kingdom. There are four major groups of fungi.

Fungi		
Group	**Examples**	**Characteristics**
Threadlike fungi	bread mold	spherical; decomposers
Sac fungi	yeast, morels	saclike; parasites and decomposers
Club fungi	mushrooms, rusts, smuts	club-shaped; parasites and decomposers
Lichens	British soldier	symbiotic with algae

Kingdom Plantae

The organisms in this kingdom are multicellular eukaryotes. They have specialized organ systems for different life processes. They are classified in divisions instead of phyla.

Plants		
Group	**Examples**	**Characteristics**
Bryophytes	mosses, liverworts	reproduce by spores
Club mosses	*Lycopodium*, ground pine	reproduce by spores
Horsetails	rushes	reproduce by spores
Ferns	spleenworts, sensitive fern	reproduce by spores
Conifers	pines, spruces, firs	reproduce by seeds; cones
Cycads	*Zamia*	reproduce by seeds
Gnetophytes	*Welwitschia*	reproduce by seeds
Ginkgoes	*Ginkgo*	reproduce by seeds
Angiosperms	all flowering plants	reproduce by seeds; flowers

Kingdom Animalia

This kingdom contains multicellular eukaryotes. They have specialized tissues and complex organ systems.

Animals		
Group	**Examples**	**Characteristics**
Sponges	glass sponges	no symmetry or segmentation; aquatic
Cnidarians	jellyfish, coral	radial symmetry; aquatic
Flatworms	planaria, tapeworms, flukes	bilateral symmetry; organ systems
Roundworms	*Trichina*, hookworms	bilateral symmetry; organ systems
Annelids	earthworms, leeches	bilateral symmetry; organ systems
Mollusks	snails, octopuses	bilateral symmetry; organ systems
Echinoderms	sea stars, sand dollars	radial symmetry; organ systems
Arthropods	insects, spiders, lobsters	bilateral symmetry; organ systems
Chordates	fish, amphibians, reptiles, birds, mammals	bilateral symmetry; complex organ systems

Using the Microscope

Parts of the Compound Light Microscope

- The **ocular lens** magnifies the image 10×.

- The **low-power objective** magnifies the image 10×.

- The **high-power objective** magnifies the image either 40× or 43×.

- The **revolving nosepiece** holds the objectives and can be turned to change from one magnification to the other.

- The **body tube** maintains the correct distance between the ocular lens and objectives.

- The **coarse-adjustment knob** moves the body tube up and down to allow focusing of the image.

- The **fine-adjustment knob** moves the body tube slightly to bring the image into sharper focus.

- The **stage** supports a slide.

- **Stage clips** hold the slide in place for viewing.

- The **diaphragm** controls the amount of light coming through the stage.

- The light source provides a **light** for viewing the slide.

- The **arm** supports the body tube.

- The **base** supports the microscope.

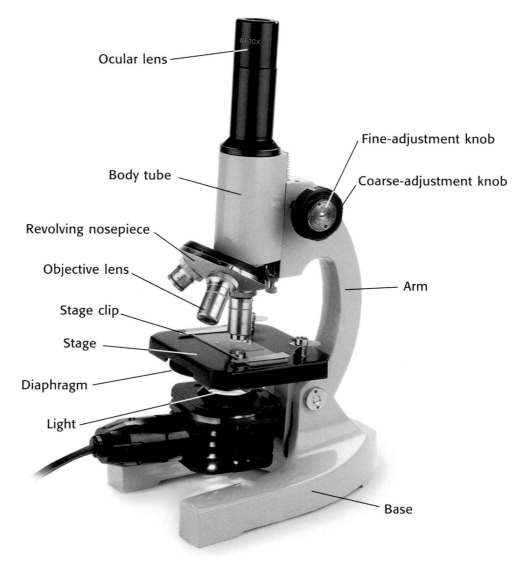

Ocular lens

Fine-adjustment knob

Body tube

Coarse-adjustment knob

Revolving nosepiece

Objective lens

Arm

Stage clip

Stage

Diaphragm

Light

Base

Proper Use of the Compound Light Microscope

1 Carry the microscope to your lab table using both hands. Place one hand beneath the base, and use the other hand to hold the arm of the microscope. Hold the microscope close to your body while moving it to your lab table.

2 Place the microscope on the lab table at least 5 cm from the edge of the table.

3 Check to see what type of light source is used by your microscope. If the microscope has a lamp, plug it in, making sure that the cord is out of the way. If the microscope has a mirror, adjust it to reflect light through the hole in the stage.
Caution: If your microscope has a mirror, do not use direct sunlight as a light source. Direct sunlight can damage your eyes.

4 Always begin work with the low-power objective in line with the body tube. Adjust the revolving nosepiece.

5 Place a prepared slide over the hole in the stage. Secure the slide with the stage clips.

6 Look through the ocular lens. Move the diaphragm to adjust the amount of light coming through the stage.

7 Look at the stage from eye level. Slowly turn the coarse adjustment to lower the objective until it almost touches the slide. Do not allow the objective to touch the slide.

8 Look through the ocular lens. Turn the coarse adjustment to raise the low-power objective until the image is in focus. Always focus by raising the objective away from the slide. *Never focus the objective downward.* Use the fine adjustment to sharpen the focus. Keep both eyes open while viewing a slide.

9 Make sure that the image is exactly in the center of your field of vision. Then switch to the high-power objective. Focus the image, using only the fine adjustment. *Never use the coarse adjustment at high power.*

10 When you are finished using the microscope, remove the slide. Clean the ocular lens and objective lenses with lens paper. Return the microscope to its storage area. Remember, you should use both hands to carry the microscope.

Making a Wet Mount

1 Use lens paper to clean a glass slide and a coverslip.

2 Place the specimen you wish to observe in the center of the slide.

3 Using a medicine dropper, place one drop of water on the specimen.

4 Hold the coverslip at the edge of the water and at a 45° angle to the slide. Make sure that the water runs along the edge of the coverslip.

5 Lower the coverslip slowly to avoid trapping air bubbles.

6 Water might evaporate from the slide as you work. Add more water to keep the specimen fresh. Place the tip of the medicine dropper next to the edge of the coverslip. Add a drop of water. (You can also use this method to add stain or solutions to a wet mount.) Remove excess water from the slide by using the corner of a paper towel as a blotter. Do not lift the coverslip to add or remove water.

Glossary

A

algae (AL JEE) protists that convert the sun's energy into food through photosynthesis (48)

angiosperm (AN jee oh SPUHRM) a plant that produces seeds in flowers (77)

Animalia the classification kingdom containing complex, multicellular organisms that lack cell walls, are usually able to move around, and possess nervous systems that help them be aware of and react to their surroundings (157)

antibiotic a substance used to kill or slow the growth of bacteria or other microorganisms (30)

Archaebacteria (AHR kee bak TIR ee uh) a classification kingdom containing bacteria that thrive in extreme environments (24, 28)

asexual reproduction reproduction in which a single parent produces offspring that are genetically identical to the parent (6)

ATP adenosine triphosphate; the molecule that provides energy for a cell's activities (13)

B

bacteria extremely small, single-celled organisms without a nucleus; prokaryotic cells (24)

binary fission the simple cell division in which one cell splits into two; used by bacteria (25)

bioremediation (BIE oh ri MEE dee AY shun) the use of bacteria and other microorganisms to change pollutants in soil and water into harmless chemicals (30)

budding a type of asexual reproduction in which a small part of the parent's body develops into an independent organism (60)

C

carbohydrate a biochemical composed of one or more simple sugars bonded together that is used to provide and store energy (11)

cell a membrane-covered structure that contains all of the materials necessary for life (4)

cell membrane a phospholipid layer that covers a cell's surface and acts as a barrier between the inside of a cell and the cell's environment (12)

cell wall a structure that surrounds the cell membrane of some cells and provides strength and support to the cell membrane (75)

cellular respiration the process of producing ATP in the cell from oxygen and glucose; releases carbon dioxide and water (111)

chlorophyll a green pigment in chloroplasts that absorbs light energy for photosynthesis (74, 110)

chloroplast an organelle found in plant and algae cells where photosynthesis occurs (51)

club fungus a type of fungus characterized by umbrella-shaped mushrooms (61)

consumer an organism that eats producers or other organisms for energy (8, 27)

controlled experiment an experiment that tests only one factor at a time (145)

cotyledon (KAHT uh LEED uhn) a seed leaf inside a seed (87)

cuticle a waxy layer that coats the surface of stems, leaves, and other plant parts exposed to air (74)

D

deciduous describes trees with leaves that change color in autumn and fall off in winter (116)

decomposer an organism that gets energy by breaking down the remains of dead organisms or animal wastes and consuming or absorbing the nutrients (8)

DNA deoxyribonucleic (dee AHKS ee RIE boh noo KLEE ik) acid; hereditary material that controls all the activities of a cell, contains the information to make new cells, and provides instructions for making proteins (6)

dormant describes an inactive state of a seed (108)

E

endospore a bacterium surrounded by a thick, protective membrane (25)

enzyme a protein that makes it possible for certain chemical reactions to occur quickly (10)

epidermis the outermost layer of cells covering roots, stems, leaves, and flower parts (89)

Eubacteria (YOO bak TIR ee uh) a classification kingdom containing mostly free-living bacteria found in many varied environments (27)

eukaryotic cell (eukaryote) (yoo KER ee OHT) a cell that contains a central nucleus and a complicated internal structure (46)

evergreen describes trees that keep their leaves year-round (116)

F

fibrous root a type of root in which there are several roots of the same size that spread out from the base of the stem (89)

Fungi a kingdom of complex organisms that obtain food by breaking down other substances in their surroundings and absorbing the nutrients (57)

fungus an organism in the kingdom Fungi (57)

funguslike protist a protist that obtains its food from dead organic matter or from the body of another organism (47)

G

gametophyte (guh MEET oh FIET) a stage in a plant life cycle during which eggs and sperm are produced (75)

gravitropism (GRAV i TROH PIZ uhm) a change in the growth of a plant in response to gravity (114)

gymnosperm (JIM noh SPUHRM) a plant that produces seeds but not flowers (77)

H

hemoglobin (HEE moh GLOH bin) the protein in red blood cells that attaches to oxygen so that oxygen can be carried through the body (10)

heredity the passing of traits from parent to off-spring (6)

homeostasis (HOH mee OH STAY sis) the maintenance of a stable internal environment (5)

host an organism on which a parasite lives (33, 48)

hyphae (HIE fee) chains of cells that make up multicellular fungi (58)

hypothesis a possible explanation or answer to a question (144)

I

imperfect fungus a fungus that does not fit into other standard groups of fungi (62)

K

kingdom the most general of the seven levels of classification (156)

L

lactic-acid bacteria bacteria that digest the milk sugar lactose and convert it into lactic acid (31)

lichen the combination of a fungus and an alga that grows intertwined and exists in a symbiotic relationship (63)

lipid a type of biochemical, including fats and oils, that does not dissolve in water; lipids store energy and make up cell membranes (12)

M

mass the amount of matter that something is made of; its value does not change with the object's location (141)

metabolism (muh TAB uh LIZ uhm) the combined chemical processes that occur in a cell or living organism (6)

meter the basic unit of length in the SI system (141)

mold shapeless, fuzzy fungi (59)

mycelium (mie SEE lee uhm) a twisted mass of fungal hyphae that have grown together (58)

N

nitrogen cycle the movement of nitrogen from the nonliving environment into living organisms and back again (29)

nitrogen fixation the process of changing nitrogen gas into forms that plants can use (29)

nonvascular plant a plant that depends on the processes of diffusion and osmosis to move materials from one part of the plant to another (76)

nucleic acid a biochemical that stores information needed to build proteins and other nucleic acids; made up of subunits called nucleotides (13)

nucleotide a subunit of DNA consisting of a sugar, a phosphate, and one of four nitrogenous bases (13)

O

ovary in flowers, the structure containing ovules that will develop into fruit following fertilization (95)

P

parasite an organism that feeds on another living creature, usually without killing it (48)

pathogenic bacteria bacteria that invade a host organism and obtain the nutrients they need from the host's cells (32)

petals the often colorful structures on a flower that are usually involved in attracting pollinators (94)

phloem (FLOH EM) a specialized plant tissue that transports sugar molecules from one part of the plant to another (88)

phospholipid a type of lipid molecule that forms much of a cell's membrane (12)

photosynthesis (FOHT oh SIN thuh sis) the process by which plants capture light energy from the sun and convert it into sugar (74)

phototropism a change in the growth of a plant in response to light (113)

phytoplankton (FITE oh PLANK tuhn) a microscopic photosynthetic organism that floats near the surface of the ocean (19)

pistils the female reproductive structures in a flower that consist of a stigma, a style, and an ovary (95)

Plantae the kingdom that contains plants—complex, multicellular organisms that are usually green and use the sun's energy to make sugar by photosynthesis (157)

pollen the dustlike particles that carry the male gametophyte of seed plants (82)

pollination the transfer of pollen to the female cone in conifers or to the stigma in angiosperms (85)

producer organisms that make their own food, usually by using the energy from sunlight to make sugar (8, 27)

prokaryotic cell (prokaryote) (proh KER ee OHT) a cell that does not have a nucleus or any other membrane-covered organelles; also called a bacterium (24)

protein a biochemical that is composed of amino acids; its functions include regulating chemical reactions, transporting and storing materials, and providing support (10)

protist an organism that belongs to the kingdom Protista (46)

Protista a kingdom of eukaryotic single-celled or simple, multicellular organisms; kingdom Protista contains all eukaryotes that are not plants, animals, or fungi (46, 156)

protozoa animal-like protists that are single-celled consumers (52)

pseudopodia (SOO doh POH dee uh) structures that amoebas use to move around (52)

R

rhizoids small, hairlike threads of cells that help hold nonvascular plants in place (78)

rhizome the underground stem of a fern (80)

S

sac fungus a type of fungus that reproduces using spores, which develop in a sac called an ascus (60)

scientific method a series of steps that scientists use to answer questions and solve problems (144)

sepals the leaflike structures that cover and protect an immature flower (94)

sexual reproduction reproduction in which two sex cells join to form a zygote; sexual reproduction produces offspring that share characteristics of both parents (6)

spore a small reproductive cell protected by a thick wall (58)

sporophyte (SPOH roh FIET) a stage in a plant life cycle during which spores are produced (75)

stamen the male reproductive structure in a flower that consists of a filament topped by a pollen-producing anther (95)

stigma the flower part that is located at the tip of the pistil (95)

stimulus anything that affects the activity of an organism, organ, or tissue (5, 113)

stomata openings in the epidermis and cuticle of a leaf that allow carbon dioxide to enter the leaf (93, 112)

symbiosis (SIM bie OH sis) a close, long-term association between two or more species (53)

T

taproot a type of root that consists of one main root that grows downward, with many smaller branch roots coming out of it (89)

threadlike fungus a fungus that develops from a spore called a zygospore (59)

transpiration the loss of water from plant leaves through openings called stomata (112)

tropism a change in the growth of a plant in response to a stimulus (113)

V

variable a factor in a controlled experiment that changes (145)

vascular plant a plant that has specialized tissues called xylem and phloem, which move materials from one part of the plant to another (77)

virus a microscopic particle that invades a cell and often destroys it (33)

volume the amount of space that something occupies or the amount of space that something contains (153)

X

xylem (ZIE luhm) a specialized plant tissue that transports water and minerals from one part of the plant to another (88)

Index

Credits

Abbreviations used: (t) top, (c) center, (b) bottom, (l) left, (r) right, (bkgd) background

ILLUSTRATIONS

All illustrations, unless otherwise noted below by Holt, Rinehart and Winston.

Scope and Sequence: T11, Paul DiMare, T13, Dan Stuckenschneider/Uhl Studios, Inc.

Chapter One Page 7 (c), Will Nelson/Sweet Reps; 12 (b), Morgan-Cain & Associates; 12 (cl), Blake Thornton/Rita Marie; 13 (tr), David Merrell/Suzanne Craig Represents Inc.; 13 (cr), John White/The Neis Group; 13 (br), Morgan-Cain & Associates; 17 (cr), Blake Thornton/Rita Marie; 18 (bl), Morgan-Cain & Associates.

Chapter Two Page 25 (cl), Art and Science, Inc.; 26, Kip Carter; 29 (bl), Carlyn Iverson; 34 (cl), Morgan-Cain & Associates; 34 (c), Art and Science, Inc.; 34 (bl), Morgan-Cain & Associates; 34 (bc), Morgan-Cain & Associates; 41 (tr), Art and Science, Inc.

Chapter Three Page 51 (c), Scott Thorn Barrows/The Neis Group; 52 (c), Scott Thorn Barrows/The Neis Group; 54, Morgan-Cain & Associates; 55, Art and Science, Inc.; 58, Will Nelson/Sweet Reps.

Chapter Four Page 75 (tr), Morgan-Cain & Associates; 75 (bl), Sidney Jablonski; 77 (c), John White/The Neis Group; 78 (art), Ponde & Giles; 78 (arrows), Sidney Jablonski; 80 (art), Ponde & Giles; 80 (arrows), Sidney Jablonski; 83 (tr), Keith Locke; 83 (cl), Sarah Woods; 83 (br), James Gritz/Photonicia; 85 (art), Will Nelson/Sweet Reps; 85 (arrows), Sidney Jablonski; 87 (art), John White/The Neis Group; 87 (arrows), Sidney Jablonski; 88 (bl), Will Nelson/Sweet Reps; 89 (tr), John White/The Neis Group; 91 (r), Will Nelson/Sweet Reps; 93-94 (tr), Will Nelson/Sweet Reps; 98 (br), Sarah Woods; 101, Will Nelson/Sweet Reps.

Chapter Five Page106 (b), Will Nelson/Sweet Reps; 107 (c) Will Nelson/Sweet Reps; 108 (b), Will Nelson/Sweet Reps; 110 (c), Stephen Durke/Washington Artists; 111 (cl), Ponde & Giles; 112 (c), Morgan-Cain & Associates; 113 (bl), Carlyn Iverson; 115 (tr), Stephen Durke/Washington Artists; 115 (bl), Rob Schuster/Hankins and Tegenborg; 117 (c), Rob Schuster/Hankins and Tegenborg; 122 (cr), Will Nelson/Sweet Reps; 123 (cr), Carlyn Iverson.

LabBook Page 135 (br), 136 (br), Keith Locke.

Appendix Page 142 (t), Terry Guyer; 146 (b), Mark Mille/Sharon Langley; 800-801.

PHOTOGRAPHY

Cover and Title page: Visuals Unlimited/Robert W. Domm

Table of Contents: Page iv(t), CNRI/Science Photo Library/Photo Researchers, Inc.; iv(b), Robert Brons/BPS/Stone; v(t), David M. Phillips/Visuals Unlimited; v(c), Ken W. Davis/Tom Stack & Associates; v(b), Sam Dudgeon/HRW Photo; vi(t), Michael Fogden/DRK Photo; vi(c), Dwight R. Kuhn; vi(b), W. Cody/Westlight/Corbis; vii(t), David Phillips/Visuals Unlimited; vii(c), Dr. Norman R. Pace & Dr. Esther R. Angert; vii(b), Oliver Meckes/MPI-Tubingen/Photo Researchers, Inc.

Scope and Sequence: T8(l), Lee F. Snyder/Photo Researchers, Inc.; T8(r), Stephen Dalton/Photo Researchers, Inc.; T10, E. R. Degginger/Color-Pic, Inc., T12(l), Rob Matheson/The Stock Market

Master Materials List: T26(br), Image ©2001 PhotoDisc; T27(bl, cr), Image ©2001 PhotoDisc

Feature Borders: Unless otherwise noted below, all images copyright ©2001 PhotoDisc/HRW. "Across the Sciences" 71, all images by HRW; "Careers" 103, sand bkgd and Saturn, Corbis Images; DNA, Morgan Cain & Associates; scuba gear, ©1997 Radlund & Associates for Artville; "Eye on the Environment" 125, clouds and sea in bkgd, HRW; bkgd grass, red eyed frog, Corbis Images; hawks, pelican, Animals Animals/Earth Scenes; rat, Visuals Unlimited/John Grelach; endangered flower, Dan Suzio/Photo Researchers, Inc.; "Health Watch" 43, dumbbell, Sam Dudgeon/HRW Photo; aloe vera, EKG, Victoria Smith/HRW Photo; basketball, ©1997 Radlund & Associates for Artville; shoes, bubbles, Greg Geisler; "Scientific Debate" 20, Sam Dudgeon/HRW Photo; "Science Fiction" 21, saucers, Ian Christopher/Greg Geisler; book, HRW; bkgd telescope, Dave Cutler Studio, Inc./SIS; "Science Technology and Society" 42, 70, 102, robot, Greg Geisler; "Weird Science" 124, mite, David Burder/Stone; atom balls, J/B Woolsey Associates; walking stick, turtle, EclectiCollection.

Chapter One: pp. 2-3 Rick Friedman/Black Star Publishing/Picture Quest; 2 Dexter Sear/IO Vision; 3 HRW Photo; 4(tr), Visuals Unlimited/Cabisco; 4(bl), Visuals Unlimited/Science Visuals Unlimited; 4(br), Wolfgang Kaehler/Liaison International; 5(cl, cr), David M. Dennis/Tom Stack and Associates; 5(br), Visuals Unlimited/Fred Rohde; 6(tl), Visuals Unlimited/Stanley Flegler; 6(c), James M. McCann/Photo Researchers, Inc.; 6(bl), Lawrence Migdale/Photo Researchers, Inc.; 8 Robert Dunne/Photo Researchers, Inc.; 9(tr), Wolfgang Bayer; 9(cr), Visuals Unlimited/Rob Simpson; 10(bl), Photo Researchers, Inc.; 10(bc), Hans Reinhard/Bruce Coleman, Inc.; 16 Visuals Unlimited/Stanley Flegler; 17 Wolfgang Bayer; 19(bl), Dede Gilman/Unicorn Stock Photos; 20(tc, c), NASA.

Chapter Two: pp. 22-23 CAMR/A. B. Dowsett/Science Photo Library/Photo Researchers, Inc.; 23 HRW Photo; 24(c), Robert Yin/Corbis; 24(cr), Dr. Norman R. Pace and Dr. Esther R. Angert; 25(cr), Institut Pasteur/CNRI/Phototake; 25(br), Heather Angel; 26(bl), Visuals Unlimited/David M. Phillips; 26(cl), Fran Heyl Associates; 26(br), CNRI/Science Photo Library/Photo Researchers, Inc.; 27(tr), SuperStock; 27(b), Larry Ulrich/DRK Photo; 28 Richard T. Nowitz/ Corbis; 30(cr), Bio-Logic Remediation LTD; 30(tl), Sergio Purtell/FOCA; 32(tl), R. Sheridan/Ancient Art & Architecture Collection; 32(bl), Visuals Unlimited/Sherman Thomson; 33(b), E.O.S./Gelderblom/Photo Researchers, Inc.; 34(cl), Visuals Unlimited/Hans Gelderblom; 34(cr), Visuals Unlimited/K. G. Murti; 34(bl), Dr. O. Bradfute/Peter Arnold; 34(br), Oliver Meckes/MPI-Tubingen/Photo Researchers, Inc.; 38(b), Institut Pasteur/CNRI/Phototake; 38(c), Robert Yin/ Corbis; 38(cr), Dr. Norman R. Pace and Dr. Esther R. Angert; 40 Fran Heyl Associates; 42 HRW Photo composite; 43 Oliver Meckes/MPI-Tubingen/Photo Researchers, Inc.

Chapter Three: pp. 44-45 Steve Taylor/Stone; 45 HRW Photo; 46(tr), Visuals Unlimited/David Phillips; 46(bl), Matt Meadows/ Peter Arnold; 46(bc), Breck Kent; 46(c), Michael Abbey/Photo Researchers, Inc.; 47(cr), David M. Dennis/ Tom Stack; 47(b), Matt Meadows/Peter Arnold; 48(tl), Dr. Bruce Kendrick; 48(bl, br), Dr. E. R. Degginger; 49(cr), Kenneth W. Fink/Photo Researchers, Inc.; 49(br), Manfred Kage/Peter Arnold; 50(tl), Robert Brons/BPS/Stone; 50(b), Kevin Schafer/Peter Arnold; 52(bl) Parks/OSF/ Animals Animals; 53(tr), Manfred Kage/Peter Arnold; 53(br), George H. Harrison/Grant Heilman; 53(bc), Dr. Hossler/Custom Medical Stock Photo; 56(cl), Dr. Hilda Canter-Lund; 56(c), Eric Grave/Science Source/Photo Researchers, Inc.; 57(cr), Runk/Schoenberger/Grant Heilman; 57(cl), Visuals Unlimited/Stan Flegler; 57(bl), David M. Dennis/ Tom Stack; 57(c), R. Carr/Bruce Coleman; 58(bl) A. Davies/Bruce Coleman; 59(tr), Ralph Eague/Photo Researchers, Inc.; 59(b), Andrew Syred/Science Photo Library/Photo Researchers, Inc.; 60(tl), Gamma-Liaison; 60(c), J. Forsdyke/Gene Cox/Science Photo Library/Photo Researchers, Inc.; 60(b), Laurie Campbell/NHPA; 61(tr), Visuals Unlimited/Wally Eberhart; 61(c, cl), Dr. E. R. Degginger; 62(tl), Michael Fogden/ DRK; 62(c), Visuals Unlimited/Inga Spence; 62(b), Walter H. Hodge/Peter Arnold; 63(c), John Gerlach/DRK Photo; 63(tr), Visuals Unlimited/Walt Anderson; 63(cr), Visuals Unlimited/Gerald & Buff Corsi; 66(cr) Visuals Unlimited/David Phillips; 67 David M. Dennis/Tom Stack; 68 David M. Dennis/ Tom Stack; 69(all), Omikron/Photo Researchers, Inc.; 70(all), Paul F. Hamlyn, BTTG; 71 Dr. James A. Pisarowicz/Wind Cave National Park.

Chapter Four: pp 72-73 Gary Braasch/CORBIS; 72 Wolfgand Kaehler/CORBIS; 73 HRW Photo; 74(cl), Robert Shafer/Stone; 74(bl, tr), SuperStock; 76(tc), Runk/Schoenberger/Grant Heilman; 76(bl), John Gerlach/Earth Scenes; 76(tl), Bruce Coleman, Inc.; 79(tr), Runk/Schoenberger/Grant Heilman; 79(bl), John Weinstein/The Field Museum, Chicago, IL; 80(tl), Larry Ulrich/DRK Photo; 80(c), SuperStock; 81(tr), Ed Reschke/Peter Arnold; 81(cr), Runk/Schoenberger/Grant Heilman; 82(tr), Robert Barclay/Grant Heilman; 82(cl), Heather Angel; 82(br), Phil Degginger; 84(tl), Tom Bean; 84(tc), Jim Strawser/Grant Heilman; 84(br), Visuals Unlimited/John D. Cunningham; 84(bl), Walter H. Hodge/Peter Arnold; 85(tr) Patti Murray/Earth Scenes; 86(tl), William E. Ferguson; 86(bl), Werner H. Muller/Peter Arnold; 86(bc), Grant Heilman; 86(br), SuperStock; 89(bc), Runk/Schoenberger/ Grant Heilman; 89(br), Nigel Cattlin/Holt Studios International/Photo Researchers, Inc.; 89(bl), Dwight R. Kuhn; 89(tr), Ed Reschke/Peter Arnold, Inc.; 89(cr), Runk/Rannels/Grant Heilman Photography; 90(tc), Harry Smith Collection; 90(cl), Larry Ulrich/DRK Photo; 90(bc), Albert Visage/Peter Arnold; 90(br), Dale E. Boyer/Photo Researchers, Inc.; 91(tl), Stephen J. Krasemann/Photo Researchers, Inc.; 91(tr), Tom Bean; 92(tr), Index Stock Photography; 92(tl, bc), Dr. E.R. Degginger; 92(cl), Gary B. Braasch; 93(br) Ken W. Davis/Tom Stack; 94(tl), SuperStock; 95(tr), George Bernard/ Science Photo Library/Photo Researchers, Inc.; 95(c), Patrick Jones/Corbis; 98(c) The Field Museum, Chicago, IL; 99 SuperStock; 100(br) Kevin Adams/ Liaison International; 102 Carl Redmond/University of Kentucky; 103(tl), Mark Philbrick/Brigham Young University; 103(br), Phillip-Lorca DiCorcia.

Chapter Five: pp. 104-105 Breck P. Kent/Animals Animals; 105 HRW Photo; 108(tl), Visuals Unlimited/W. Ormerod; 108(tc), George Bernard/Earth Scenes; 108(tr), Image Copyright ©2001 Photodisc, Inc.; 109(tl), Paul Hein/Unicorn; 109(cr), Jerome Wexler/Photo Researchers, Inc.; 109(cl), George Bernard/Earth Scenes; 110(br), Gregg Hadel/Stone; 112(tl), Dr. Jeremy Burgess/Science Photo Library/Photo Researchers, Inc.; 113(br), Cathlyn Melloan/Stone; 114(cl, cr), R. F. Evert; 115(c), Dick Keen/Unicorn; 115(b), Visuals Unlimited/E. Webber; 116(t), W. Cody/WestLight; 116(bl, bc, br), Rich Iwasaki/Stone; 117(tl), Visuals Unlimited/ Bill Beatty; 117(cl), Visuals Unlimited/Bill Beatty; 121(cl) R. F. Evert; 121(tr, cr, br), Rich Iwasaki/Stone; 123(b), W. Cody/Westlight; 124(cl), Discover Syndication/Walt Disney Publications; 124(bc), David Littschwager & Susan Middleton/Discover Magazine; 124(br), Discover Syndication/Walt Disney Publications; 125(cr), Cary S. Wolinsky.

Labook: "LabBook Header": "L", Corbis Images, "a", Letraset-Phototone, "b" and "B", HRW, "o" and "k", Images Copyright ©2001 PhotoDisc, Inc.; 118-119, Sam Dudgeon/HRW Photo; 127(cl), Michelle Bridwell/HRW Photo; 127(br), Image Copyright ©2001 Photodisc, Inc.; 128(cl) Victoria Smith/HRW Photo; 128(bl), Stephanie Morris/HRW Photo; 129(tr), Jana Birchum/HRW Photo; 137(bc), Breck P. Kent; 137(br), Stephen J. Krasemann/ Photo Researchers, Inc.; 137(c), Visuals Unlimited/R. Calentine; 137(t), Runk/ Schoenberger/Grant Heilman.

Appendix: p. 158 CENCO

Sam Dudgeon/HRW Photos: all Systems of the Body background photos, viii-1, 10(br), 11, 14, 19(tr, cr, br), 110(bl), 126, 127(b); 128(tr, br), 129(tl), 130, 131, 133, 135, 138, 143(br).

Peter Van Steen/HRW Photos: p. 15, 31(tr, bl), 36, 41, 65, 129(b), 143(tr).

John Langford/HRW Photos: p. 127(tr).

Acknowledgements continued from page iii.

Alyson Mike
Science Teacher
East Valley Middle School
East Helena, Montana

Donna Norwood
Science Teacher and Dept. Chair
Monroe Middle School
Charlotte, North Carolina

James B. Pulley
Former Science Teacher
Liberty High School
Liberty, Missouri

Terry J. Rakes
Science Teacher
Elmwood Junior High School
Rogers, Arkansas

Elizabeth Rustad
Science Teacher
Crane Middle School
Yuma, Arizona

Debra A. Sampson
Science Teacher
Booker T. Washington Middle School
Elgin, Texas

Charles Schindler
Curriculum Advisor
San Bernadino City Unified Schools
San Bernadino, California

Bert J. Sherwood
Science Teacher
Socorro Middle School
El Paso, Texas

Patricia McFarlane Soto
Science Teacher and Dept. Chair
G. W. Carver Middle School
Miami, Florida

David M. Sparks
Science Teacher
Redwater Junior High School
Redwater, Texas

Elizabeth Truax
Science Teacher
Lewiston-Porter Central School
Lewiston, New York

Ivora Washington
Science Teacher and Dept. Chair
Hyattsville Middle School
Washington, D.C.

Elsie N. Waynes
Science Teacher and Dept. Chair
R. H. Terrell Junior High School
Washington, D.C.

Nancy Wesorick
Science and Math Teacher
Sunset Middle School
Longmont, Colorado

Alexis S. Wright
Middle School Science Coordinator
Rye Country Day School
Rye, New York

John Zambo
Science Teacher
E. Ustach Middle School
Modesto, California

Gordon Zibelman
Science Teacher
Drexell Hill Middle School
Drexell Hill, Pennsylvania

Self-Check Answers

Chapter 1—It's Alive!! Or, Is It?

Page 5: Your alarm clock is a stimulus. It rings, and you respond by shutting it off and getting out of bed.

Chapter 2—Bacteria and Viruses

Page 27: Cyanobacteria were once classified as plants because they use photosynthesis to make food.

Chapter 3—Protists and Fungi

Page 48: No, some funguslike protists are parasites or consumers.

Page 54: 1. Cilia are used to move a ciliate through the water and to sweep food toward the organism. 2. Ciliates are classified as animal-like protists because they are consumers and they move.

Page 58: 1. Both are consumers that secrete digestive juices onto a food source and then absorb the digested nutrients. Both reproduce asexually by spores. 2. Hyphae grow together to form the mycelium.

Page 59: Spores are contained in sporangia.

Chapter 4—Introduction to Plants

Page 76: Plants need a cuticle to keep the leaves from drying out. Algae grow in a wet environment, so they do not need a cuticle.

Page 92: Stems hold up the leaves so that the leaves can get adequate sunshine for photosynthesis.

Chapter 5—Plant Processes

Page 107: Fruit develops from the ovary, so the flower can have only one fruit. Seeds develop from the ovules, so there should be six seeds.

Page 111: The sun is the source of the energy in sugar.

Page 114: 1. (See concept map below.)
2. During negative phototropism, the plant would grow away from the stimulus (light), so it would be bending to the left.

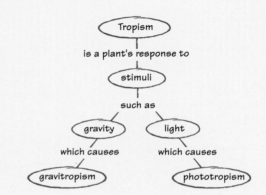